BEBOP

Bebop
The Music and Its Players

Thomas Owens

New York Oxford
OXFORD UNIVERSITY PRESS
1995

Oxford University Press

Oxford New York Toronto
Delhi Bombay Calcutta Madras Karachi
Kuala Lumpur Singapore Hong Kong Tokyo
Nairobi Dar es Salaam Cape Town
Melbourne Auckland

and associated companies in
Berlin Ibadan

Published by Oxford University Press, Inc.,
200 Madison Avenue, New York, New York 10016

Oxford is a registered trademark of Oxford University Press

Library of Congress Cataloging-in-Publication Data
Owens, Thomas, 1938–
Bebop : the music and its players / Thomas Owens.
p. cm. Includes bibliographical references and index.
ISBN 0-19-505287-0
1. Bop (Music)—History and criticism.
2. Jazz musicians—United States.
I. Title.
ML3506.095 1994 781.65′5—dc20 93-32504

9 8 7 6 5 4 3 2 1

Printed in the United States of America
on acid free paper

Permission to reprint excerpts from listed musical compositions is gratefully acknowledged:

MPL Communications, Inc.

"Autumn Leaves (Les Feuilles Mortes)," English lyric by Johnny Mercer, French lyric by Jacques Prevert, music by Joseph Kosma. Copyright © 1947, 1950 (Renewed) Enoch et Cie. Sole Selling Agent for U.S. & Canada: Morley Music Co. by Agreement with Enoch et Cie. All Rights Reserved. Used By Permission.

"Woodyn' You," music by Dizzy Gillespie. Copyright © 1943 (Renewed) Edwin H. Morris & Company, a Division of MPL Communications, Inc. All Rights Reserved. Used By Permission.

Music Sales Corporation

"Darn That Dream" by Eddie DeLange and Jimmy Van Heusen. Copyright © 1939 (Renewed) by Lewis Music/Music Sales Corporation (ASCAP) and Scarsdale Music. International Copyright Secured. All Rights Reserved. Reprinted By Permission.

"How Could You Do a Thing Like That to Me," words by Allan Roberts, music by Tyree Glenn. Copyright © 1955 (Renewed) by Music Sales Corporation (ASCAP) and Allan Roberts Music Co. International Copyright Secured. All Rights Reserved. Reprinted By Permission.

"There Is No Greater Love" by Isham Jones and Marty Symes. Copyright © 1936 (Renewed) by Music Sales Corporation (ASCAP). International Copyright Secured. All Rights Reserved. Reprinted By Permission.

"Will You Still Be Mine" by Thomas Adair and Matt Dennis. Copyright © 1940 (Renewed) by Music Sales Corporation (ASCAP). International Copyright Secured. All Rights Reserved. Reprinted By Permission.

Second Floor Music

"Daahoud" by Clifford Brown. Copyright © 1963 (Renewed 1990) Second Floor Music.

Sophisticate Music

"Boplicity" by Cleo Henry.

Warner/Chappell Music, Inc.

"All This and Heaven Too" by Max Steiner. Copyright © 1940 Warner Bros. Inc. (Renewed). All Rights Reserved. Used By Permission.

"The Fruit" by Earl Budd Powell. Copyright © 1953 (Renewed) Windswsept Pacific Entertainment Co. d/b/a Longitude Music Co. All Rights Reserved. Used By Permission.

"A Gal in Calico" by Arthur Schwartz and Leo Robin. Copyright © 1946 Warner Bros. Inc. (Renewed). All Rights Reserved. Used By Permission.

"Hot House" by Tadd Dameron. Copyright © 1954 WB Music Corp. (Renewed). All Rights Reserved. Used By Permission.

"Too Marvelous for Words" by Richard Whiting and Johnny Mercer. Copyright © 1937 Warner Bros. Inc. (Renewed). All Rights Reserved. Used By Permission.

"What Is This Thing Called Love" by Cole Porter. Copyright © 1939 Warner Bros. Inc. (Renewed). All Rights Reserved. Used By Permission.

In addition, we gratefully acknowledge permission to reprint:

Musical example from *Jazz Piano, a Jazz History* by Billy Taylor. Used by permission of Duane Music, Inc. (ASCAP).

Material from *Down Beat*, 16 (9 Sept. 1949), 22 (2 Nov. 1955), 26 (14 May 1959), 27 (29 Sept. 1960), 43 (11 March 1976), 44 (15 Dec. 1977), 47 (Jan. 1980), 59 (March 1992). Reprinted with permission of *Down Beat*.

Material from *Jazzletter* 3 (October 1983), 8 (August 1989, September 1989), 9 (November 1990). Reprinted with permission of Gene Lees.

Preface

The location is Catalina Bar and Grill, one of the most attractive jazz clubs in Los Angeles, located just a few doors away from where Shelly's Manne Hole used to be. It is Wednesday night, 10 August 1988, the second night of a six-night engagement of the Milt Jackson Quartet. Joining the famed vibraphonist are pianist Cedar Walton, bassist John Clayton, and drummer Billy Higgins. Well before the start of the 9 p.m. set the club fills with musicians and laymen. (Before the night ends, Monty Alexander, Buddy Collette, Slide Hampton, John Heard, Art Hillery, and other players will drop by to hear their colleagues.)

When the musicians enter they take their places on the band platform, looking happy and well dressed (in the 1980s, suits and ties became common jazz concert attire again, even in casual Southern California). They play one great piece after another: a standard, a ballad, a blues, a series of Thelonious Monk pieces, and so on for one and a half hours. Clearly the four men are having a wonderful time playing for and with one another, and the audience members are the grateful beneficiaries of the musicians' mutual joy. Jackson's rhythmically ingenious, bluesy melodies inspire frequent joyous outbursts from the audience. Walton's broad harmonic palette and endless supply of tongue-in-cheek song quotes heighten everyone's pleasure. Clayton's luscious, singing *arco* tone when he plays *'Round Midnight* and *Django* is awe-inspiring; toward the end of the *Django* solo Jackson shakes his head in disbelief and says "Amazing!" softly to himself. And smiling Billy Higgins (does anyone else have more fun playing jazz?), one of the world's most sensitive and telepathic drummers, shapes and guides the progress of each piece.

The set ends with the inevitable *Bags' Groove,* Jackson's best-known melody. It has been an electric set of acoustic jazz, as verified by the damp but happy faces of the players and the prolonged applause of the audience. It was bebop at its finest, played by master musicians in their prime, and played for a racially and chronologically mixed audience of enthusiastic supporters. Surely bebop was alive and well in Los Angeles in 1988!

The evening described above is but a microcosm of bebop in the late 1980s and early 1990s. Jackson is one of several pioneer players who are still active. Walton and Higgins represent the players who expanded the vocabulary in the 1950s and 1960s, and Clayton epitomizes the large group of gifted players born in the 1950s and 1960s who maintain the tradition with passion, skill, and inventiveness. Bebop, with three generations of practitioners, is a flourishing jazz style in the 1990s, just as it was in 1940s, 1950s, 1960s, 1970s, and 1980s.

Bebop, first established in the early and middle 1940s in New York City by Charlie Parker, Dizzy Gillespie, Thelonious Monk, Bud Powell, Kenny Clarke, Max Roach, Ray Brown, and a few others, is now, after half a century of evolution, a diversified jazz language, spoken by hundreds of internationally prominent musicians. This analogical reference to language is hardly new. Westerners have called music an "international language" for generations, and jazz musicians often speak of "telling a story" when they play. Barry Velleman makes this suggestion:

> Much of the enchantment of jazz results from its similarities to conventional spoken language. Although not lacking in errors and hesitations, spoken language is often more dynamic than the cautious, elaborated written form. We can compare normal conversation to jazz improvisation through its individualism, creativity, endless possibility, spontaneity, thematic continuity, and inherent imperfection. (Velleman 1978: 31)

In a similar vein, Perlman and Greenblatt argue that the harmonic and melodic constraints a jazz musician deals with are "in many ways analogous to the syntactic and semantic constraints of natural language and that playing an improvised solo is very much like speaking sentences." (Perlman & Greenblatt 1981: 169)

The analogy between music and language does not include musical equivalents for verbs, nouns, and tenses, but each musical style does have its vocabulary of rhythmic, melodic, harmonic, textural, and timbral elements that all members of an ensemble must share if they are to make excellent music within the style. One of my main goals in this book is to explicate that vocabulary as it occurs in bebop. In the course of this explication, references to Charlie Parker's personal style abound. Because of his position as the most important role model in early bebop, I have written a relatively detailed summary of his style in Chapter 3. This summary, in turn, is the basis for the discussions of many other players throughout the book.

My other main goal is to discuss the individual styles of the most important bebop players. Sometimes that goal was easy to pursue. Every jazz historian agrees that the seven bebop pioneers listed earlier (Parker, Gillespie, Monk, and so on) played vital roles in establishing the idiom. And there are younger players that also would appear on any writer's list of important beboppers. Conversely, every writer would agree that some major figures are not part of the bebop world. Thus, they assign Count Basie and Benny Carter to the swing style, though these men made many fine recordings with bebop groups. The task becomes difficult, agonizingly difficult, when selecting the bebop players to omit. There are many hundreds of first-rate bebop musicians in the world; they deserve space in a book devoted to their music. But this book is a survey, not an encyclopedia. To treat fairly George Coleman, Buddy Collette, Junior Cook, Bob Cooper, Allen Eager, Booker Ervin, Joe Farrell, Frank Foster, Benny Golson, Eddie Harris, Tubby Hayes, Jimmy Heath, Joe Henderson, Buck Hill, Red Holloway, Plas Johnson, Clifford Jordan, Richie Kamuca, Rahsaan Roland Kirk, Harold Land, Yusef Lateef, Charles Lloyd, James Moody, Oliver Nelson, Sal Nistico, Bill Perkins, Herman Riley, Ronnie Scott, Wayne Shorter, Herbie Stewart, Lew Tabackin, Grover Washington, Ernie Watts, Frank Wess, and the other tenor saxophonists that I have either omitted entirely or treated in a cursory fashion,[1] I would have to write a book just on bebop tenor saxophonists. And then there would be other books to write on alto saxophonists, trumpeters, and so on. If I have delineated a representative array of individual styles in this overview of the bebop language, then I have done what I set out to do.

Writers can discern the boundary between swing and bebop, and can point out the swing and bebop elements in early 1940s recordings, with relative ease. But the ending limit of the bebop idiom is more difficult to define, and writers have reached no general agreement on the matter. Some would restrict bebop to many, but not all, of the young players of the 1940s. On the other hand, A. B. Spellman extends the style to include Ornette Coleman and Cecil Taylor (Spellman 1966). Some say that Miles Davis was a "cool" player, not a bebop player. Others would say he was a bebopper until the late 1950s or early 1960s, when he moved into "modal," or "vamp style" jazz (Kernfeld 1981). Or perhaps his bands left bebop in the middle 1960s, with pieces close stylistically to those of Ornette Coleman, or at the end of the 1960s, with the adoption of rock elements into the musical mix. My position on these matters unfolds throughout the book. Whether you accept or reject my view that a certain piece, player, or ensemble is within the bebop world, I hope you will agree that the music and musicians I discuss are worth discussing. Jazz is a world of great beauty, and if that beauty reaches you it should not matter what label we affix to it (or, as Gertrude Stein said, ". . . a rose . . .").

This book is by no means a comprehensive survey of bebop. I have not lived long enough to have heard all the relevant recordings. Indeed, no one could possibly listen to, evaluate, and analyze everything in the vast literature of this music. Many readers doubtless will wonder why I omitted this or that recording. The omission may be a result of consciously selecting something else that I like better, or of selecting one that better suits the narrative of the moment. Or I may not know the recording in question. But usually jazz recordings represent arbitrary slices of the players' musical lives, and the jazz world attaches an importance to specific recordings that may have little or nothing to do with the players themselves. For example, Charlie Parker's *Koko* and Miles Davis's *So What* are important primarily because they were so widely imitated, not because they were intrinsically better than dozens of other recorded solos by these artists. Had I left out the discussions of those two performances my descriptions of Parker's and Davis's styles and their importance to jazz would be no different.

Some jazz history books place primary emphasis on the early recordings of far too many players. Jazz is a complex music that requires time and maturity to master. Often the best work of players comes during his/her middle or late years; Benny Carter, Duke Ellington, Johnny Hodges, Sidney Bechet, and others come to mind from earlier style periods. In the world of bebop, Dexter Gordon, Milt Jackson, Gil Evans, Hank Jones, Tommy Flanagan, Lee Konitz, and Phil Woods are but a few who have recorded their best work years after making their initial marks in the international jazz world. For example, the album Stan Getz recorded during the week preceding his 50th birthday far surpassed anything he did during his twenties. I have tried to focus upon both the later and the earlier works of these great creators.

The large central portion of the book, Chapters 3 through 9, presents players grouped according to their principal instruments. Leonard Feather (1957) and Joachim Berendt (1959) used a similar plan in their historical overviews of jazz, and Ira Gitler (1966) used it to discuss players of the 1940s. An additional model was the instrumentation of a typical bebop band: alto or tenor saxophone (Chapters 3 to 5), trumpet (Chapter 6), piano (Chapter 7), bass and drums (Chapter 8). These chapters on individual instrumentalists lead to Chapter 10, which focuses on some of the great ensemble performances of bebop. The final chapter is an overview of young bebop players, most of whom were born in the 1960s and 1970s, whose greatest work may well occur in the 21st century.

I fear that some will read this book without listening to the music under discussion. Words are dismally inadequate substitutes for the actual music. The notated examples are closer to the real thing, but are also inadequate no matter how carefully written, and are far too short and restricted to convey a sense of the complete performance. I hope that readers will use this book as a tour guide for exploring some of the greatest jazz ever recorded. Such an exploration takes dedication, time, and (unless you have some friends who collect jazz recordings) money. But the rewards go far beyond the monetary. Those of us who carry in our inner ears and memories listening experiences such as those described at the beginning of this preface and throughout the book are rich in ways that really matter.

Acknowledgments

I owe a debt of gratitude to many people for help of various kinds:

To the Owenses—Karen, Holly, Scott, and Steve—for their forbearance during the past few years;

To Sheldon Meyer, the supervising editor of this book (and many others over the years); he is a great cheerleader, an unbelievably patient man, and a gentleman;

To Joellyn Ausanka, my associate editor, for attending to the many matters, small and large, connected with getting this book into print;

To Stephanie Sakson, my copy editor, for her great skill in catching missing hyphens, misused prepositions, misplaced modifiers, and much more;

To the members of the Department of Ethnomusicology at UCLA, for offering me a visiting professorship during 1991–92; the lighter teaching load provided much needed extra time for work on this book;

To Theodore Dennis Brown, for letting me use information from his dissertation in Chapter 8;

To Betty Price at UCLA and to Roger Quadhamer and Susan Dever at El Camino College, for giving me access to institutional laser printers to generate preliminary versions of this book;

To UCLA students Fred Samia and Ethan Schwartz, for their proofreading skills;

To Mel Gilbert and Joanna Bosse of Mark of the Unicorn, for their help with Mosaic, the notation program I used for the musical examples;

To Gene Midyett, Charles Moore, and Gabe Kreisworth, for loaning some important recordings to me from their personal collections, and to Gabe for guiding me to some important players that I would have overlooked;

To Safford Chamberlain, for sharing his insights into Warne Marsh's music;

To Bill Green, for his valuable explanations of saxophone techniques and equipment;

To Bobby Shew, for his equally valuable explanations of trumpet techniques and equipment;

To John Clayton, for his generous help with bassists and bass techniques, and to Richard Simon, for his equally generous help and for filling my ears with thousands of great bass lines during hundreds of rehearsals and gigs;

To Gary Frommer, for his help with drummers and drumming techniques, and much more; during the countless jobs we have played together he has taught me more about jazz than any other person in my life. Most of his teaching was nonverbal, or rather, supra-verbal; and

To Steve Bailey, Jimmy Bunn, Kenny Burrell, Gary Burton, Benny Carter, Buddy Collette, Chick Corea, Eddie Daniels, Richard Duffy, Mercer Ellington, Clare Fischer, Jimmy Ford, Frank Foster, Victor Gaskin, Dizzy Gillespie, Joe Hackney, Jeff Hamilton, Bob Hammer, Lionel Hampton, Jimmy, Percy, and Tootie Heath, Woody Herman, David Hughes, Milt Jackson, Connie Kay, John Kirkwood, Jimmy Knepper, Mike Lang, Bruce Lett, John Lewis, Don and Jeff Littleton, Bobby McFerrin, Branford and Wynton Marsalis, Llew Matthews, Kevin O'Neal, John Patitucci, Tim Pope, Tito Puente, Jimmy Rowles, Howard Rumsey, Paul Smith, Richard Stoltzman, Billy Taylor, Bobby Thomas, Frank Tiberi, John Ward, Mike Whited, Gerald Wilson, and Dave Young, all of whom have answered my questions in classes and workshops at El Camino and UCLA, or in informal conversations.

A special thank you goes to Horace Silver, a teaching colleague and friend as well as one of my musical heroes. I have learned a great deal about jazz, and life, from this kind man.

Finally, my thanks to all the great musicians—those listed above and many others—whose music has enriched my life and inspired me to write this book.

Torrance, California T.O.
May 1994

Contents

BEBOP

CHAPTER ONE

The Beginnings

"Bebop" was a label that certain journalists later gave it, but we never labeled the music. It was just modern music, we would call it. We wouldn't call it anything, really, just music. (Kenny Clarke[1])

In the Onyx Club [1944], we played a lot of original tunes that didn't have titles. . . . I'd say, "Dee-da-pa-n-de-bop . . ." and we'd go into it. People, when they'd wanna ask for one of those numbers and didn't know the name, would ask for bebop. And the press picked it up and started calling it bebop. (Dizzy Gillespie[2])

[Bebop is] trying to play clean and looking for the pretty notes. (Charlie Parker[3])

The word "bop" is a most inadequate word. . . . (Ralph Ellison[4])

Let's not call it bebop. Let's call it music. (Charlie Parker[5])

Parker's suggestion in that last quote is understandable, but modern man seldom settles for such broad, generalized categories. We subdivide things, then subdivide the subdivisions. Thus jazz, already a subdivision of African-American music, has its early jazz, classic jazz, New Orleans jazz, dixieland, Chicago dixieland, swing, prebop, bebop (also known as rebop or bop), cool, progressive jazz, West Coast jazz, East Coast jazz, hard bop, funky jazz, free jazz, freebop, fusion, and more. Admittedly, several of these terms are synonyms, but jazz does have its subcategories, and one of them is bebop. Although it is a silly-sounding name, originating from linguistically meaningless vocables used by jazz musicians,[6] it is here to stay as the label for a primary jazz style.

When bebop was new, many jazz musicians and most of the jazz audience heard it as radical, chaotic, bewildering music. But time and familiarity softened and even eliminated the objections. The first-generation beboppers became senior citizens enjoying worldwide acclaim. They inspired many great bebop musicians worldwide, musicians young enough to be their children and grandchildren. Each generation of players has added to this musical vocabulary without altering its basic syntax. Bebop, in fact, is now the *lingua franca* of jazz, serving as the principal musical language of thousands of jazz musicians. It also affects the way earlier jazz styles are played, and is the parent language of many action jazz ("free jazz") and fusion players, who from time to time return to the idiom, not to parody it but to honor it and to reaffirm their musical roots.

During the 1930s and early 1940s the predominant jazz style was swing. Swing, a dance-oriented style, typically was played in ballrooms by big bands of fourteen or more musicians (though small swing groups played in clubs). Swing rhythm sections provided a solid, basic accompaniment, built largely of long quarter-note strings embellished by the standard high-hat pattern:  . Swing harmonies centered around mildly dissonant chords such as dominant sevenths, dominant ninths, and major and minor triads with added sixths. These harmonies supported the largely diatonic melodies of riff-dominated jazz tunes and of popular songs. These themes most often were either 12-measure blues songs or 32-measure melodies in *aaba* or *abac* form. The arrangement often was more interesting than the theme, for arrangers ingeniously altered the dull rhythms of popular songs, passed around the melody among the trumpet, trombone, and sax sections, created background patterns (often riffs) to be played during solo passages, and added original introductions, interludes, and codas to dress up the choruses. Often the arrangements left only short and discontinuous passages for solo improvisations, though Count Basie's band and a few others regularly emphasized solos more than written passages, especially in public performances.

The first bebop recordings, which appeared in the mid 1940s, offered striking contrasts to the norms of the swing style. The music, mostly played by small groups (quartets, quintets, sextets), was generally more complex. In particular, bebop rhythm sections, using varied on- and off-beat chordal punctuations (known as comping) supplied by pianists and guitarists, and additional punctuations supplied by the drummer on drums and cymbals, were much more polyrhythmic than were swing rhythm sections. Also, bebop harmonies were more dissonant and the melodies improvised upon these harmonies were more chromatic and less symmetrical rhythmically. Bebop themes, while mostly 12-measure blues and 32-measure melodies in *aaba* form (just as in swing), often were more chromatic, disjunct, and rhythmically complicated. On the other hand, bebop arrangements were simpler than the often intricate arrangements of swing bands. The typical bebop arrangement consisted of an improvised piano introduction, a theme chorus (played in unison by groups containing two or three wind instruments), a series of improvised solo choruses, and a closing theme statement. This simple design gave maximum emphasis to improvised melodies and to the rhythmic and harmonic interplay between soloist and rhythm section.

In the late 1930s and early 1940s swing was extremely popular in the United States; many commercial recording artists of the time were either jazz musicians themselves or had close ties with the jazz world. The American public was familiar and comfortable with big-band swing. But the complexities of bebop bewildered most of this audience. When the first bebop recordings came out in 1944–45, the new music seemed to have come from nowhere. A national strike by the American Federation of Musicians against the recording industry from 1942 to 1944 precluded the recording of performances that might have shown the new style gradually evolving from the old.[7] Today we can look back and recognize traces of this evolution in the studio recordings made before and immediately after the strike. Also, some private recordings made during that period and released on LP years later supply some missing evolutionary links. Bebop clearly had a firm foundation in the swing style that preceded it; several musical components are common to both styles.

One swing element that bebop adopted was the tritone substitution—in essence, replacing the V^7 chord with a dominant seventh on scale-degree $\flat 2$, a tritone (three whole steps) away. Thus, in the key of C, the fundamental progression G^7-C becomes $D^{\flat 7}$-C. (Both chords in these progressions commonly contain one or two enriching notes such as the ninth, the augmented eleventh, or the thirteenth.) Duke Ellington built this substitution into some of his arrangements. Pianist Art Tatum used it in his florid reharmonizations of popular standard songs, and saxophonist Coleman Hawkins, trumpeter Roy Eldridge, and clarinetist Benny Goodman acknowledged it in their improvised melodies, most often by playing the melodic line $\flat 9$-8 or $\flat 9$-7-8. Bebop players regarded V^7 and $\flat II^7$ as interchangeable, and used the $\flat 9$-7-8 figure repeatedly. Soon they applied this little chromatic encircling figure to the third and fifth of the chord as well (see Charlie Parker's Fig. 5A in Chapter 3, p. 32).

Several other melodic figures came to bebop from swing. One occasionally used by Hawkins became a common phrase beginning—a rising triplet arpeggio preceded by a lower neighbor (Fig. 1A in Chapter 3). Saxophonist Lester Young also used the rising arpeggio but reconfigured it rhythmically so that the goal of arrival was lower than the highest note of the arpeggio (Fig. 1B in Chapter 3). Another pattern was a simple ornament, the inverted mordent (Fig. 2A in Chapter 3). Hawkins and Young used (and Young eventually abused) this simple figure. Short chromatic ascents and descents consisting of two, three, or four half-steps (Figs. 4A–F in Chapter 3) appeared in solos by Hawkins, Young, Eldridge, guitarist Charlie Christian, and others. Occasionally swing players used a figure involving a rising dominant-minor-ninth arpeggio (Fig. 3 in Chapter 3), and a four-note chromatic encircling figure (Fig. 5B in Chapter 3), both of which became extremely common in the bebop vocabulary.

Vibratos are usually slower in bebop than in swing; on the average, bebop vibratos are about 4.7 oscillations per second, swing vibratos about 5.6 oscillations per second.[8] Lester Young, who beboppers admired greatly, was probably the direct inspiration for this slowing. His vibrato speed—about 4.5 oscillations per second—may in turn derive from saxophonist Frankie Trumbauer's (about 4.8 oscillations per second), Young's early role model.

Young also was almost certainly the inspiration for the "beat-stretching" approach to ballads favored by bebop players beginning in the mid-1940s. For a few beats at a time Young would lag slightly behind the beat set by the rhythm section, setting up a subtle rhythmic tension. It is an effective device when done well. But it can backfire on the player when overdone, as some young players showed in their early attempts to emulate him, and as Young himself proved in his last recordings.

The harmonic vocabulary of bebop has a higher level of dissonance than that of swing. Bebop players did not add any dissonant chords to jazz at first, however; they merely used them more frequently than had their predecessors. One need only listen to Ellington's extraordinary *Ko-Ko* or to Tatum's *Body and Soul* to realize that the raw materials of bebop harmony were already in place in jazz, ready to be reapplied to the newer idiom.[9]

The bebop "revolution" (a common but suspect description) was primarily rhythmic; bebop rhythm sections produce a more complex, multilayered texture than their swing-era counterparts. One obvious change was the more varied rhythmic role of pianists. Swing pianists often played a striding left hand (imitating the "oom-pah" or "boom-chick" of march accompaniments) and melodies in the right hand while soloing, and a "boom-chick" pattern divided between the two hands when accompanying. But Ellington, Count Basie, and Nat Cole sometimes played irregularly timed chordal punctuations while accompanying their wind players; that is, they "comped" before comping was the norm in jazz. And Basie pioneered in the thin-textured solo style where the left hand plays soft sustained chords or occasional punctuations ("comping" again) instead of the automatic and full accompaniments of the ragtime and stride pianists. Basie's solo style, while built around simple chords and lean melodies, laid the foundation for bebop solo piano texture. In the process he also lightened the texture of the rhythm section, making it easier to hear the bassist's lines.

Most swing-style bassists emphasized roots and fifths of chords, or chordal arpeggios, played one note per beat (i.e., |C-C-G-G-|D-D-G-G-|, or |C-E-G-A-|C-A-G-E-|).[10] But Walter Page in the Basie band and Jimmy Blanton in the Ellington band regularly "walked"; that is, they emphasized stepwise motion, which became the norm for bebop bassists. Further, when Blanton played solos—and bass solos were rare in the swing era until this young giant appeared—he played melodic lines reminiscent of wind players, not of bass players. That is, he did not worry about touching on roots of chords, but instead floated his lines above some imaginary bass line. Finally, he found ways of playing pizzicato bass lines in a more legato manner. His tone quality, legato walking bass lines, and solo style were models for the best of the early bebop bassists.[11]

Bebop drumming was truly a revolution (and a revelation, when recording technology improved enough to capture it properly). The multiple rhythmic layers that many bebop drummers regularly produce differ markedly from the more monorhythmic norms of swing drumming. Yet some bebop rhythms came from earlier drummers. For example, Jo Jones, the great drummer in the Basie band, made exemplary use of the high hat. Not content to play endlessly the standard high-hat pattern—[musical notation]—he created a repertory of small but important variations on it that bebop drummers adapted to the ride cymbal. He, and other swing drummers, occasionally would punctuate the music with snare-drum strokes, rim shots, or bass-drum strokes, a practice extended greatly by bebop drummers.

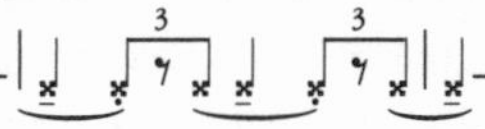

The most important contributions that swing made to bebop, however, were the players themselves. The bebop pioneers, we must remember, grew up listening to swing, and when they began playing they played swing. Dizzy Gillespie's early recordings show no trace of the new idiom; he was a swing player heavily influenced by Roy Eldridge. Charlie Parker's early recordings also reveal a swing style player, albeit one with a distinctive personality. Similarly, Kenny Clarke and other members-to-be of the jazz avant-garde had their early professional experiences as swing-band musicians. They learned the technique, time, tone, and improvisation of jazz within the swing tradition. Then gradually, and at first with little or no intent of altering jazz, they modified their manner of playing. Eventually, inevitably, there was a new way to perform improvised African-American instrumental music.

Tracing the individual player's musical evolution from swing to bebop is not always easy; for several players it is impossible, since they began recording only after assimilating the new idiom. The clearest trail to follow is Dizzy Gillespie's (see pp. 101–4), for most of the essential recorded landmarks in his early career appear on one beautifully produced anthology, *Dizzy Gillespie: The Development of an American Artist, 1940–1946* (Smithsonian Collection R 004-P4-13455), issued in 1976. The lengthiest samples of Gillespie's early style were recorded at Minton's Playhouse, one of two small clubs in New York City's Harlem—the other was Monroe's Uptown House—that served as focal points for the young players formulating the new style. Both clubs featured after-hours jam sessions that attracted traveling and resident musicians. Besides Gillespie, the frequent participants at these sessions were trumpeters Roy Eldridge, Joe Guy (who was in the house band at Minton's), John Carisi, and Hot Lips Page, alto saxophonist Charlie Parker, tenor saxophonist Don Byas, pianists Kenny Kersey, Thelonious Monk (the house pianist at Minton's), Al Tinney (the house pianist at Monroe's), and Mary Lou Williams, guitarist Charlie Christian (who kept an extra amplifier at Minton's to use whenever he arrived in town), bassist Nick Fenton (the house bassist at Minton's), and drummer Kenny Clarke (the house drummer at Minton's).

According to contemporary reports by those present, the music at these sessions was small-group swing at first, but gradually changed to something else. Jazz fan Jerry Newman made some recordings on his portable disc recorder in 1941 that provide an imperfect glimpse back into those proceedings. Years later some of these informal recordings were issued commercially.[12] Their fidelity is poor and the rhythm sections are seriously undermiked. But they are all we have, for none of these ad hoc groups made any studio recordings.

The music in these Minton's recordings is swing, not bebop. Yet Gillespie's solos reveal a player searching for a new way to express himself. Still largely an Eldridge disciple, he nonetheless surprises us here and there with phrases that are foreign to the language of swing. As it turns out they are foreign to his future bebop language, as well. In the example below, from the beginning of his second chorus in *Kerouac* (based on the chords of *Exactly like You*), measures 2 through 6 are jarringly unconvincing, surrounded as they are by the tamest of phrases.

In a few years he learned to integrate augmented elevenths and altered dominant chords more smoothly into his melodic fabric.[13]

Charlie Parker's early style, in contrast to Gillespie's, seems to have derived from several sources, no one of them primary. He was a great admirer of Lester Young, some of whose recorded solos he learned note-for-note. A snippet of Young's famous 1936 recording of *Oh, Lady Be Good!* appears in Parker's 1940 recording of the same piece,[14] and a few other traces of Young's music (such as the phrase labeled 2B in Chapter 3) appear in the recordings Parker made in 1940 and 1942 with the Jay McShann band. He also admired alto saxophonist Buster Smith; the two were sidemen in a local band in Kansas City in the late 1930s. Sitting next to Smith night after night apparently had its effect on Parker, for his solo in *Moten Swing* is remarkably similar in tone quality and some melodic details to Smith's solo recorded three weeks earlier.[15] And he obviously listened to Coleman Hawkins, a bit of whose famous *Body and Soul* recording he quotes in his solo on the same piece.[16] In the private recordings he made with Gillespie and Oscar Pettiford (p. 102), he produces a tone quality on the tenor saxophone that is a curious blend of Hawkins's and Young's timbres.

Ultimately the more lasting swing traces in Parker's style were a set of melodic formulae. The patterns such as those labeled Figures 1A, 1B, 2A, 3, 4A–F, 5A, and 5B on pp. 31–32 were in the air during the late 1930s and early 1940s. Parker must have learned them from several sources, and applied them in his unique way.

The key drummer in these early years of experimentation was Kenny Clarke, who played with Gillespie in Teddy Hill's big band and then served as house drummer at Minton's. He preferred to play the standard high-hat pattern on the ride (or top) cymbal. In addition, he stopped automatically playing the bass drum on every beat and began playing a variety of on- and off-beat punctuations on the bass drum and snare drum ("bombs").[17] Moving his right hand from the high hat (situated on his left) to his ride cymbal (on his right) gave him more room to maneuver his left hand on the snare drum (directly in front). He abandoned the automatic four-to-the-measure bass-drum habit unexpectedly:

> Well, I never expected . . . things to turn out the way they did as far as modern drumming is concerned . . . It just happened . . . accidentally. . . . What really brought the whole thing about, we were playing a real fast tune once . . . with Teddy [Hill]. . . . The tempo was too fast to play four beats to the measure, so I began to cut the time up . . . But to keep the same rhythm going . . . I had to do it with my hand, because . . . my foot just wouldn't do it. So I started doing it with my hand, and then every once in a while I would kind of lift myself with my foot, to kind of boot myself into it, and that made the whole thing move. So I began to work on that. . . .
>
> I played a long time with Diz [Dizzy Gillespie], from 1938 to 1942. Then I left for the Army, and during my absence, Dizzy, who plays the drums well, taught all the other drummers my way of playing.

These rhythmic changes are hard to hear on records, for drums were served poorly by the microphones and recording machines of the time. On most recordings one can hear only a fraction of what Jo Jones, Kenny Clarke, and others played. Bass drums and ride cymbals are all but inaudible, as are the subtler strokes on high hat and snare drum. Perhaps the best recorded sample of Clarke's work during the transition years is with a quintet led by Sidney Bechet. In the final exuberant choruses of *One O'Clock Jump*,[18] his off-beat bass-drum punctuations and ride cymbal patterns are unusually clear for recordings at the time. He is not playing bebop, but neither is he playing in the traditional New Orleans or swing styles. It is a fascinating moment. A year later the microphone for the recording machine in Minton's was near the drum set during the piece *Swing to Bop*,[19] and captured a rare sampling of his richly varied and highly interactive playing on snare and bass drums.

By 1944 a handful of young players had worked out their new musical vocabulary and were performing regularly, mostly in the small nightclubs on 52nd Street in New York City. During the same year the American Federation of Musicians settled their strike against the recording companies. And so the stage was set for the first recordings of the new jazz.

CHAPTER TWO

Early Classics

In the winter of 1943–44, the first small group organized to perform the still-unnamed new music debuted at the Onyx Club on 52nd Street in New York City. It was a quartet, then a quintet, co-led by Dizzy Gillespie and bassist Oscar Pettiford. The other members were pianist George Wallington, drummer Max Roach, and then tenor saxophonist Don Byas.[1] The group remained intact only a few months and did not record as a quintet, though some of the members recorded as sidemen for other leaders. Pettiford's solo on Coleman Hawkins recording of *The Man I Love* was the first bebop bass solo on record.[2] Two months later, Gillespie, Pettiford, and Roach recorded six pieces with Hawkins's twelve-man ensemble, a big band without trombones (see below). And in April, Gillespie and Pettiford were part of Billy Eckstine's big band when it made its recording debut. Some melodically adventurous unison parts for Eckstine's trumpet section in *I Stay in the Mood for You* pointed toward the style the band was soon to follow.[3]

There were a few other recorded samples of the new style. In January, pianist Bud Powell recorded eight pieces with a sextet led by Cootie Williams and four more with Williams's big band. He comped effectively, and in *Floogie Boo, I Don't Know,* and *Honeysuckle Rose* he played brief solos using the texture that was to become the norm for bebop pianists: fluid, single-note melodies dominated by strings of two and three-eighth notes per beat in the right hand, and unpredictably timed chordal punctuations in the left.[4] Later that year he played a jarringly modern two-chorus solo in the otherwise conservative performance by Williams's big band of *Royal Garden Blues.*[5] And in September, Charlie Parker recorded with a quintet led by guitarist-singer Tiny Grimes (see below). None of these early recordings truly represents bebop, primarily because the young players were outnumbered sidemen and not leaders. But they do help to paint a sketchy picture of how the new style emerged.

The real bebop revolution (or evolution, as Gillespie preferred to say) was in the rhythm section. Recording techniques of the 1940s still captured only partially the sound of the string bass and drum set, so aural evidence of early changes in rhythm-section playing is skimpy. Art Blakey's playing in *Blowing the Blues Away* and *Opus X* by the Eckstine band is one of the least murky recordings of early bebop drumming.[6] During part of each piece Blakey's ride-cymbal pattern and snare-drum fills project reasonably well and, combined with pianist John Malachi's modest amount of comping, provide an early recorded sample of a bebop rhythm section.

The early recordings preserve the repertoire better than the performance techniques; one example is Gillespie's piece *Woody 'n' You.* He first recorded it with Coleman Hawkins in 1944, and soon it became a bebop standard.[7] It is 32 measures long, subdivided into *aaba* form—the most common form used in the 1930s and 1940s for popular songs. The *a* section consists of three two-measure sequences on ii-V chords, ending on the tonic (D♭); its harmonic plan is

$\mathrm{G{\small MI}}^{7(\flat 5)}$—$\mathrm{C}^{7(\#9)}$—$\mathrm{F{\small MI}}^{7(\flat 5)}$—$\mathrm{B}^{\flat 7(\#9)}$—$\mathrm{E}^{\flat}\mathrm{{\small MI}}^{7(\flat 5)}$—$\mathrm{A}^{\flat 7(\#9)}$—$\mathrm{D}^{\flat}\mathrm{{\small MA}}^{9}$
$[\mathrm{ii}^{ø7}$ — $\mathrm{V}^{\#9}]\mathrm{iii}$ $[\mathrm{ii}^{ø7}$ — $\mathrm{V}^{\#9}]\mathrm{ii}$ $\mathrm{ii}^{ø7}$ — $\mathrm{V}^{\#9}$ — I^{9}

The alternation between the half-diminished ii chords (one of Gillespie's favorite harmonic sonorities) and the dominant seventh-raised ninth V chords is a pungent mix of dissonances that appealed to the young modernists. Also appealing was the little figure involving scale steps ♭9-7-8 (Fig. 5A on page 32, common to both styles) at the end of each *a* section. The only other bebop touch in this recording, however, was Gillespie's solo, with several of his favorite melodic gestures (see pp. 102–3). The forward-looking rhythm section (Pettiford, Roach, and pianist Clyde Hart) is too poorly recorded to reveal much beyond basic time-keeping. The dominating sonority in the record is the warm, vibrato-laden sound of swing-style saxophones.

Similar comments on style apply to Parker's only recording session of 1944. He was the lone bebop player in Tiny Grimes's quintet when Grimes recorded *Tiny's Tempo* (a blues), *Red Cross* (based on the chords of *I Got Rhythm*), and two pieces that feature Grimes's singing.[8] Two aspects of this session are important to the early history of bebop. First, there are multiple takes of each of the four pieces, which provide a new perspective on Parker's skill as an improviser. His three different solos on *Tiny's Tempo* and two different solos on *Red Cross* are largely different from one another, showing that his command of the idiom was thorough and secure. He had no need for a worked-out, memorized solo. Second, the two all-instrumental pieces were prophetic, for the blues and the *I Got Rhythm* harmonic structures became the most common harmonic plans in bebop during the late 1940s. Entire recording dates by some leaders (not Gillespie or Parker) were devoted to these two harmonic structures.

Further progress in documenting the new style came in early 1945, mostly in small group sessions led by Gillespie. First, Gillespie's sextet recorded four pieces, the most forward-looking being Tadd Dameron's *Good Bait.*[9] Here, besides Gillespie's modern solo lines, are Clyde Hart's creditable comping, Pettiford's short bebop solo, and an inventive and complex second theme played in unison in the final chorus. In the other three pieces the stylistic balance still favors swing. Nonetheless, *Salt Peanuts* is the first recording of Gillespie's humorous classic, *Be-Bop* officially acknowledges the new name for the new music; and *I Can't Get Started* shows us Gillespie's important reharmonization of this popular song (discussed on page 103).[10]

In January 1945, Gillespie recorded several pieces with Boyd Raeburn's big band, which was exploring bebop in the mid-1940s. The most important result of this temporary partnership was the Latin-tinged *A Night in Tunisia,*[11] a piece that Gillespie had written in 1942 and arranged for Earl Hines's big band. Cast in the usual *aaba* mold, its most ear-catching features are its chord-derived main motive—

—and its insistent use of the ♭II-i harmonic relationship (E♭13-DMI), the tritone substitution. The *a* section ends with V^7-i, but the E♭-D relationship moves into the melody, in the form of the ♭9-7-8 figure that Gillespie brought with him from his Eldridge-inspired early style. Once the theme ends, the band abandons the Latin rhythms and plays a 12-measure interlude that has been integral to the piece ever since. Gillespie's flashy two-bar solo break leads into his solo chorus, during which we can hear Shelly Manne's ride-cymbal pattern and snare drum fills. There will be better recordings of Gillespie's fine piece in future years, among them septet versions by Gillespie and by Parker, recorded a year later.[12] But this one, the parent recording of the rest, is a respectable beginning.

In February, Gillespie led two different sextet recordings of *Groovin' High.*[13] The second is the better of the two, and is the first famous bebop recording. Here performing together are the two leading bebop spokesmen, Gillespie and Parker. Granted, their rhythm section, except for the timidly comping pianist Clyde Hart, is swing-oriented. But the nature of the theme, the primary soloists, and the fame of this piece make this an early bebop classic.

First, the theme. Filled with explicit outlines of the ii-V progression in three different keys and of the iii^7-$^\flat iii^7$-ii^7 progression, it is significantly more complex than the typical swing-era theme. It has a stronger inner logic, derived from the frequent pairing of a simple two-note figure and a two-measure string of eighth notes:

The *Groovin' High* melody, an excellent and popular theme from the early years, is also a prime example of a common compositional procedure in early bebop. Gillespie built this engaging melody (officially he wrote it with Kirby Stone) to fit with the chords of *Whispering,* a popular song of the 1920s. By 1945 the use of a pre-existent chord structure was hardly new; there are many pieces in the swing repertory based on the chords of *Tiger Rag, I Got Rhythm,* and other earlier pieces. (For that matter, *someone* must have been the first to write the 12-measure blues chord structure, the basis for thousands of compositions.) But at that time Gillespie's melodic contrafact[14] was the most complex jazz melody superimposed on a pre-existing chordal scheme.

The arrangement is atypically elaborate for bebop performances, with its composed six-measure introduction (most bebop introductions are improvised and run four or eight measures), its modulations (most bebop pieces stay in one key), its choruses of varying lengths, and its dramatic half-speed coda. The performance is overly segmented because the soloists have only a half-chorus each; they play well, but they have too little time to express themselves.

Two other pieces emerged from this second *Groovin' High* session:

All the Things You Are and *Dizzy Atmosphere.* The first is a fragmented and lackluster performance of Jerome Kern's popular song. Its most lasting feature is its eight-measure introduction and coda, built upon two chords:

| D♭MI7 | D♭MI7 | C7(#9) | C7(#9) | D♭MI7 | D♭MI7 | C7(#9) | C7(#9) |

Gillespie derived this introduction from his arrangement of *Good Jelly Blues,* which he wrote for Billy Eckstine's big band.[15] Ever since this early 1945 recording session, Gillespie's introduction has been a permanent feature of jazz performances of *All the Things You Are. Dizzy Atmosphere* is more spirited, and features excellent full-chorus solos by Parker and Gillespie. The rhythm section even approximates the bebop style, with drummer Cozy Cole playing the ride-cymbal pattern and some well-intentioned snare-drum fills. The theme is a simple and repetitive tune in *aaba* form, with the *a* section based on the *a* of *I Got Rhythm* and the *b* section based on chromatically descending dominant sevenths. But the nicest surprise of the recording is, as in *Good Bait* a year earlier, a second theme that Parker and Gillespie play in unison during most of the last chorus. Unlike the riff-filled main theme, it is inventive and varied, matching more closely the melodic styles used when these men improvised.

The next landmark record date occurred in May, when Gillespie and Parker recorded the definitive versions of *Salt Peanuts* and *Hot House.* For these pieces they used the model bebop quintet instrumentation: trumpet, saxophone, piano, bass, and drums. Four of the players played important roles in defining and consolidating the new style: Parker, Gillespie, pianist Al Haig, and bassist Curley Russell. Only drummer Sid Catlett is of the older school. Still, he adapts creditably to bebop on *Hot House,* and his drive and musicality are strong enough on *Salt Peanuts* to override the stylistic discrepancies.

Both themes are melodic contrafacts: *Salt Peanuts* is based on the chords of *I Got Rhythm,* and *Hot House* on Cole Porter's *What Is This Thing Called Love?* The melodies of these two pieces in *aaba* form represent opposite tendencies in bebop themes. *Salt Peanuts* is simplicity itself—a four-measure riff phrase played twice in each *a* section, and a slightly more complex bridge (which incorporates the ubiquitous ♭9-7-8 figure twice). The main riff is really in swing style; perhaps that is why Catlett's drumming fits the piece so well.[16] *Hot House,* on the other hand, is the most chromatic and angular melody that bebop had produced by that date. But in the complexity of overall design, these two performances switch roles. *Hot House* follows the simple norm in bebop—introduction, unison statement of theme, solo choruses, unison statement of theme. *Salt Peanuts,* on the other hand, is a full-blown arrangement:

Introduction (16 measures)—8-measure drum solo, ending with Catlett playing the rhythm of the main theme; then 8 measures by the quintet
Chorus 1 (32 measures)—unison theme statement
Interlude (8 measures)—unison phrase in the more complex bebop melodic style
Chorus 2 (32 measures)—theme, with Parker playing the first two measures of the riff and Gillespie comically singing "salt PEA-nuts, salt PEA-nuts" in response; Parker improves in the bridge
Interlude 2 (16 measures)—8-measure solo by Haig; then another complex unison phrase, then a two-measure solo break by Haig
Chorus 3 (32 measures)—Haig's solo
Chorus 4 (32 measures)—Parker's solo
Interlude 3 (10 measures)—6-measure dialogue between Parker and Gillespie; then a flashy 4-measure solo break by Gillespie
Chorus 5 (32 measures)—Gillespie's solo
Half-chorus 6 (16 measures)—Catlett's solo
Coda (16 measures)—a reprise of the introduction, with everyone singing the "salt PEA-nuts" figure at the end

Throughout the piece the quintet plays well; the composed sections are precise and clean, driven by Catlett's conservative but impeccable drumming. Al Haig's solo is competent if uneventful, while Parker's is a model of a finely crafted solo and Gillespie's is a typically exuberant emotional high point.

In *Hot House* we get to hear all the main ingredients of bebop assembled in one recorded performance for the first time. After a brief drum introduction, the complex theme begins, with its chromatic melody hopping around within chords such as $C^{13(\sharp 11)}$ and $\mathrm{FMI}^{9(\mathrm{MA}7)}$. Complementing this angular melody is Haig's energetic comping and Catlett's punctuating fills, all supported by Russell's relentless walking bass lines. This multilayered texture continues during Parker's and Gillespie's solo choruses and the concluding theme chorus. All that is missing is the more aggressive drum punctuations and the driving ride-cymbal pattern that Kenny Clarke, Max Roach, or Art Blakey would have provided. Parker and Gillespie create excellent samples of their rhythmically varied improvising styles. Both scatter unexpected accents here and there in their eighth-note-dominated melodies. Both interrupt these eighth-note passages with dazzling sixteenth-note phrases that are as expertly wrought as the phrases in which they have twice the time to think.

Running through the performance is a unifying thread, derived from the opening notes of the hidden theme:

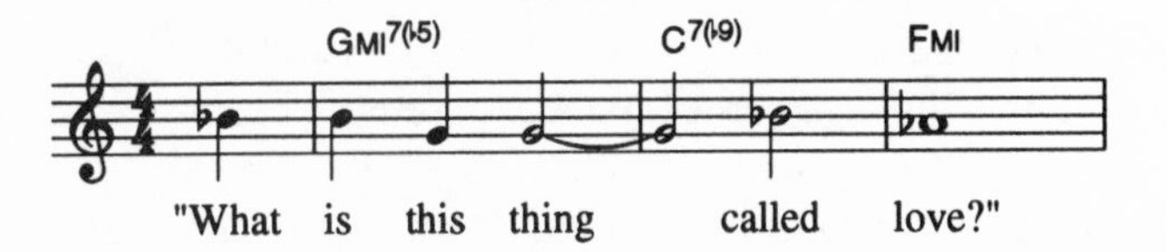

The *Hot House* melody begins with an elaboration of this phrase:

In measure 9 of his solo, Parker plays his own elaboration:

Then he begins the bridge with another variant, transposed up a fourth to fit the new key of the moment:

Finally, he uses this last variant to begin the final four measures of his solo:

Then Gillespie, as though taking the baton in a relay race, opens his solo chorus with the same figure. The final leg of the relay race is the return of the *Hot House* theme (which is in *abca* form, although it fits a harmonic scheme in *aaba* form).

On 26 November 1945, after several recording sessions as a sideman, Charlie Parker had his first session as leader. Out of this session came two famous blues melodies entitled *Now's the Time* and *Billie's Bounce,* an *I Got Rhythm* contrafact entitled *Thriving from a Riff* (later to be altered and renamed *Anthropology*), and a monument called *Koko.*[17] For the first three of these pieces Parker had the young and inexperienced Miles Davis as his front-line partner (see page 114), a rhythm team of Curley Russell and Max Roach, and pianists Dizzy Gillespie (an above-average comper) and Argonne Thornton. The best of these three is *Billie's Bounce.* It follows the typical bebop blues plan of piano introduction, two unison theme choruses, solo choruses, and two more unison theme choruses. Parker plays a fine four-chorus solo as expected, Davis is adequate, and Max Roach's fully idiomatic bebop percussion work is recorded well (although the jazz world had to await the compact disc and sophisticated remastering techniques to hear how well).

The fast *Koko* (♩ = 300) was an amazing and perplexing recording when first released. Today it stands as a classic of the jazz tradition. It begins with an unusually long introduction, 32 measures, played by Parker, Roach and Gillespie, the latter replacing Davis on trumpet. This introduction has four subsections of eight measures each: a unison melody, an improvisation by Gillespie, another by Parker, and another unison melody. Next come two masterfully executed improvised choruses by Parker on the chords of *Cherokee,* Ray Noble's popular song of the late 1930s. Each chorus is 64 measures long, on a chord structure in *aa'ba'* form. Roach follows with a half-chorus solo (32 measures). Then Parker, Gillespie, and Roach repeat the introduction, but stop abruptly three measures early, producing an enigmatic ending on the dominant.

The story behind *Koko* is convoluted. In night-club engagements Parker and Gillespie had been playing a fast arrangement of *Cherokee.* Miles Davis did not know the arrangement, with its complicated introduction/coda, so when the players decided to record it, Gillespie took over on trumpet. After the introduction on take 1, Gillespie began to play the *Cherokee* melody, while Parker played a harmony part for four measures and then an improvised counterpoint for a few more measures. But soon he stopped playing and halted the recording. Evidently he felt that with the lengthy introduction and one chorus devoted to the theme, there would be little time left for his solo. So for take 2, the famous take, he discarded the theme, leaving space for the two solo choruses that saxophonists have been copying ever since (see also pp. 38–40). Since there is no trace of the *Cherokee* melody, and since Parker had collected $300 from the owner of the Savoy for four original compositions, the record producer Teddy Reig gave to this performance the name *Koko,* and used it to complete the set of four originals.

There is a question concerning who plays piano on *Koko.* According to Reig, Gillespie played piano during Parker's two choruses and trumpet during the introduction and coda. That explanation is plausible, but during the few extant measures of the *Cherokee* melody in take 1 someone plays piano while Gillespie continues on trumpet. That pianist could have been Davis, who had a moderate command of the keyboard, or Argonne Thornton, a marginally competent pianist who had already played on *Thriving from a Riff* (unfortunately). So any of three people could have comped for Parker on take 2.[18] The issue is incidental, however, for the comping is basic, functional, and entirely subservient to the stunning solo that Parker created. It would have mattered little if the entire performance had taken place without a piano player.

By the end of 1945 a repertory of new pieces was on record and in use publicly by an ever-increasing number of polished players. In December, Gillespie, Parker, Haig, vibraphonist Milt Jackson, bassist Ray Brown, and drummer Stan Levey took their music to Los Angeles, where bebop already was developing among a lesser-known group of players. During 1946, Gillespie and Parker, now leading separate groups in different parts of the country, added *Confirmation, 52nd Street Theme, Anthropology, Moose the Mooche, Yardbird Suite, Ornithology, Our Delight,* and others to the recorded repertory, and made important new recordings of *'Round Midnight* and *A Night in Tunisia.* By this time Gillespie and Parker were no longer the main sources of the rapidly expanding inventory of bebop recordings. Don Byas already had begun to incorporate elements of bebop into his Coleman Hawkins-based style, and was continuing his recording career. Younger tenor saxophonists Sonny Stitt, Dexter Gordon, Wardell Gray, and Stan Getz each led recording sessions; so did Kenny Clarke, back from a three-year tour of duty in the armed services. Trombonist J. J. Johnson had his first session as leader and added some bebop flavoring to the Basie band. The swing bands of Woody Herman and Stan Kenton were evolving rapidly into bebop bands, as younger sidemen replaced their swing-oriented predecessors. And pianist Lennie Tristano, guitarist Billy Bauer, and their coterie began recording their own distinctive bebop dialect.

Ornithology, written by Benny Harris and Parker, was an important addition to the repertory in 1946. In some ways it is a perfect symbol of early bebop: a complex melodic contrafact (the hidden theme is *How High the Moon*) with an erudite title that refers to a founding father of bebop (Parker's nickname was "Bird"). The *Ornithology* melody begins with a phrase that had been in Parker's improvising vocabulary since at least 1942 (Parker's Fig. 2B in Chapter 3), when he used it to begin his solos in the Jay McShann recordings of *The Jumpin' Blues* (a.k.a. *Jump the Blues*):[19]

The original recording of the piece is from a productive session by Parker that also produced first recordings of *Moose the Mooche* and *Yardbird Suite,* and Parker's famous recording of *A Night in Tunisia.*[20] However, it is a recording of a theme in progress, for the winds abruptly stop their unison playing at the end of measure 12, leaving the pianist to fill in for four measures until the theme (in *aa′* form) begins again. Near the end of the theme chorus the winds continue playing, but they simply toss the same triplet figure from one to another. In later public performances Parker used an improved theme, one in which section *a* is complete and section *a′* has a better ending. Later performances also preserve more extended solos. But this inaugural recording is important because it is one of three takes that offer us another chance to compare Parker's musical thought processes from one moment to the next. As expected, his solos on each take are quite different from one another.

Ornithology and most other important pieces in the bebop repertoire came out of sessions by small groups (combos), because bebop is small-group jazz primarily. But almost from the beginning there have been some big bands devoted to the music. In 1944–45 the Eckstine and Raeburn bands played early roles in the music's evolution. And in 1945 Gillespie formed a big band, the first of several short-lived big bands under his aegis.

The big bands were even less well served than small groups by recording techniques of the time. But Gillespie's second band, of 1946–50, made some noteworthy additions to the repertory. One piece carries the linguistically meaningless title *Oop-Pop-a-Dah.*[21] A blues, it is a vehicle for the bebop-style scat singing of Gillespie and Kenny Hagood, and is the first of the bebop scat vocals on record. In it, *trumpeter* Gillespie, like Louis Armstrong two decades earlier, showed that he was also an important jazz *singer.* Even more than Armstrong, Gillespie infused his scat solos with ample doses of humor. But anyone who listens to the notes and rhythms and not just the funny vocal sounds will discover the same melodic ingenuity in his scat solos that exists in his trumpet solos.

Manteca was another of his best big band recordings. Like *A Night in Tunisia,* it represents Gillespie's interest in Latin rhythms.[22] He co-composed it with Cuban percussionist Chano Pozo, whom he added to his rhythm section in the fall of 1947. Pozo, playing conga drum, and bassist Al McKibbon begin the record with a rhythm based on straight eighth notes rather than on the swing eighths of 1940s jazz, and the trombones follow suit as they enter with an ostinato. The *a* section of the theme is repetitive and based on only three notes; one of those notes is scale step $^{\flat}7$ of the Mixolydian scale, a scale integral to many Cuban melodies. The hint of Mixolydian and the syncopated Afro-Cuban rhythms of the theme and its accompanying ostinato set this piece apart from most bebop of the 1940s. But Gillespie's irrepressible interest in moving harmonies surfaces in the bridge, suggesting that the embrace of the Cuban idiom is not complete. Sure enough, once the theme ends and the improvisations begin, the Cuban effect almost disappears as tenor saxophonist Big Nick Nicholas reverts to the enriched major mode and the *I Got Rhythm* chord changes of normal bebop. Gillespie and the full band continue the bebop mood, using swing eighths in spite of Pozo's continuing even eighths, until the final *a* section of the theme returns. Complete assimilation of Afro-Cuban rhythms and improvisations on a harmonic ostinato was still a few years away for the beboppers in 1947.

Perhaps the finest example of big-band bebop in the 1940s came from the Woody Herman band. Herman, whose swing-style clarinet and saxophone playing hardly changed at all during his career, was nonetheless sympathetic to the new idiom. By late 1947 his was essentially a bebop band, though its repertory included his earlier swing-style hits as well. In particular, his baritone saxophonist, Serge Chaloff, was very much a Parker imitator, and his tenor players, Herbie Stewart, Stan Getz, and Zoot Sims, all had developed a similar stylistic blend of Lester Young and Parker. The playing of these four men inspired arranger-saxophonist Jimmy Giuffre to compose a piece for them, appropriately titled *Four Brothers.*[23]

The melody of *Four Brothers* begins immediately, played in four-part harmony by the featured saxophonists. Giuffre's theme is an attractive, arpeggio-filled melody in *aaba* form. The bridge (which is harmonically identical to Benny Carter's *When Lights Are Low* for the first four measures) moves rapidly through chords and key areas. In the first six measures the players visit B major, D major, and C major before easing back into the home key of A$^{\flat}$ when the final *a* returns. After the theme chorus each soloist has a half-chorus solo. Stewart and Sims each solo in the first half of the second and third choruses, where the chords are easier; Chaloff and Getz begin their solos in the thicket of bridge chords, and have some difficulty, Chaloff more than Getz. The last two choruses are written out for the band except for a bridge in which Herman solos on clarinet. His solo sounds assured and polished, but consists almost entirely of the roots of the chords.[24] In the final chorus Don Lamond's drum fills are overstuffed, but otherwise his playing, captured with remarkable clarity by the recording engineer, is excellent and contributes greatly to the success of the performance. The piece ends with an exuberant coda in which each saxophonist has a final brief say.

By 1946–47 the basic parameters of bebop were in place. But some players made music that differed from the norm in some ways. Two of them were pianists: Thelonious Monk and Lennie Tristano. Monk, little-known and even less appreciated in the 1940s, had a much larger impact on jazz in the 1950s and 1960s, so I will examine his music in Chapter 7. Tristano was better-known to the jazz audience, and had a small school of musical disciples by the end of the 1940s; among them were alto saxophonist Lee Konitz, tenor saxophonist Warne Marsh, and guitarist Billy Bauer.

Tristano's bebop differed from mainstream bebop in three ways. First, the melodies he and his colleagues composed were more angular, less clearly related to their underlying harmonies, and less symmetrical in phrase lengths than melodies by Parker, Gillespie, Dameron, and others. Second, the harmonic textures created by his small groups, which usually included Bauer, were denser than those of most bebop groups. Both he and Bauer regularly comped energetically and with little apparent concern for staying out of each other's way. And third, he much preferred drummers to use brushes rather than sticks, even in lively pieces. To many listeners his music was intellectual, reserved, and "cool."

Subconscious-Lee illustrates his group style well.[25] The title is a play on words, the composer being Lee Konitz. Another melodic contrafact of *What Is This Thing Called Love?* (the Tristano school's favorite harmonic structure), it has even more surprising rhythms and startling melodic twists and turns than occur in *Hot House,* the earlier piece on the same structure. The emphasized chord tones in several places are highly dissonant notes, such as augmented elevenths of major and dominant seventh chords (i.e., F# in C chords) and major sevenths in minor chords (i.e., E in FMI chords). The solos by Tristano, Bauer, and Konitz that follow, however, are less adventuresome. In fact, other than an almost total lack of syncopation, their improvisations are similar to those played by most beboppers of the time. Rhythmic interest here, instead of coming from the soloists, comes from the syncopated comping of Tristano and Bauer, and especially from the buoyant and creative brush work by Shelly Manne.

Later in the year, Tristano, Konitz, Bauer, Marsh, and bassist Arnold Fishkin recorded some remarkable collective improvisations:

Intuition and *Digression*.[26] These startling pieces are not part of the jazz tradition of that time. Neither piece swings, nor was it intended to; the time sense is much closer to that of European chamber music, with very little syncopation and with subtle gradations of tempo. The harmonic vocabulary, largely an incidental by-product of unplanned but interactive polyphony, is rich and varied, with a high level of dissonance. The through-composed form of each is radically different from the usual theme and variations of jazz. These two pieces are early examples of what came to be called "free jazz" in the 1960s, although their impact on "free jazz" players such as Cecil Taylor, Ornette Coleman, and Archie Shepp was minimal or nonexistent. The pieces remain today as interesting anomalies, having only the most tenuous connection to the evolution of bebop.

In 1948 a new type of ensemble formed, originally to rehearse some new arrangements by Gil Evans, Gerry Mulligan, and John Lewis. It had an unusual instrumentation: trumpet, trombone, French horn, tuba, alto and baritone saxes, plus the usual rhythm trio. In essence it was a reduced version of the Claude Thornhill big band, which also incorporated a French horn and a tuba. In fact, part of the group was from the Thornhill band, and Gil Evans was Thornhill's arranger. Miles Davis joined the group and soon became its leader. In September, they played an engagement at the Royal Roost night club in New York City. A few months later the nonet, with shifting personnel, did a series of three recording dates. The twelve pieces from these sessions attracted much attention. Since most of the pieces were in slow or moderate tempos and played with quiet dynamics, they seemed to signal the start of the "cool era" in jazz history.

The importance of these recordings rests in the compositions, the arrangements, and the ensemble style of performance rather than in the solos, most of which are short and superficial, some of which are weak. And there are other performance flaws, chiefly in the fast pieces *Move* and *Budo*. These pieces, though well arranged by Lewis, suffer from a lack of rhythmic cohesion, for the ensemble has trouble keeping up with Max Roach. Perhaps they could not hear him well; a famous picture taken during the recording session shows his drums almost hidden from view behind some tall sound baffles, and widely separated from the bassist and everyone else. While the engineer captured his drum and cymbals sounds well, he may have forced the musicians to pay a price in the process.

Flaws aside, this set of twelve performances is a remarkable and important contribution to the music. There is much to enjoy here, such as the surprising meter shifts (almost unheard of in jazz then) in Mulligan's

Jeru, and the recurring shifts from major to minor in John Carisi's bittersweet blues *Israel.* Above all there is the blended sound of those six winds. All the arrangers used the winds effectively, but the sound that Gil Evans extracted from them in *Boplicity* is extraordinarily beautiful. Here, as in his earlier arrangements for the Thornhill band, Evans revealed some of the creativity that he would bring to the masterpieces *Miles Ahead* and *Sketches of Spain* a decade later (see pp. 120–23).[27]

The Davis nonet is the best-known of the groups that are larger than a typical combo but smaller than a big band. There were other such ensembles. Bill Harris recorded with nonets in 1946 and 1949, Dave Brubeck recorded with a nonet in 1948 and an octet in 1950, Shorty Rogers led octets and nonets in the early 1950s, Gigi Gryce wrote for and recorded with groups of both sizes in 1953, and Dave Pell led an octet during most of the 1950s. But except for the Pell octet, these groups were unable to function over an extended period. Neither fish nor fowl, they were too large for small clubs with low budgets, yet not big and loud enough to match the power of the big band in larger venues. So bebop continued to evolve without them.

Before leaving this overview of early bebop, let us look at two brief examples of ensemble playing by some of its best players. Drawn from late recordings by Charlie Parker on Verve, they are acoustically among the best recordings he made. The first shows the second eight measures of the opening theme statement of *Confirmation.*[28] It is a fine example of a rhythm section—pianist Al Haig, bassist Percy Heath, and drummer Max Roach—finding ways to enhance Parker's interpretation of the melody. The quartet probably performed from memory, for Parker's composition had been in the bebop repertory since 1946. At most there might have been a lead sheet in the studio, but lead sheets are silent on matters of chord voicings, bass lines, and rhythmic punctuations. Those issues are left for the players to settle, either spontaneously or, as is more likely with these experienced players, on the basis of previous performances.

Roach clearly knows this melody well. Besides keeping the time moving forward with his ride cymbal and high hat he plays a variety of punctuations that either reinforce important syncopations in the melody (measures 2, 3, and 4) or provide a rhythmic complement to the melody (measures 5 and 8). Haig also has two roles, to supply both harmonic background and rhythmic punctuation. He obviously knows the tune also, for his most assertive moments are during rests in the melody, when his syncopated comping is the most effective. Heath's jobs are to coordinate with the drummer in keeping the time moving forward and to coordinate with the pianist by sounding the roots of chords on downbeats (or on beats 1 and 3 if there are two chords per measure, as in measures 6 and 7). But here, he also participates in the rhythmic complement of the melody, by playing a well-timed syncopation in measure 1. Thus, the rhythm section functions as a team, with the members coordinating beautifully among themselves and with Parker. The result is a superb performance of Parker's fine theme.

There is a harmonic clash in the second measure. Heath plays A^7/E (a passing chord connecting F and DMI), while Haig uses an $E^\flat$ in his enriched A^7 chord. There is also a clash in measure 6, between the D that Parker plays and the $D^\flat$ of Haig's $E^{\flat 9}$. Such clashes are nearly inevitable in bebop, with its rich assortment of 7th, 9th, 11th, and 13th chords and chord substitutions. Players often take different harmonic paths to get from one structurally important chord to the next; the resulting momentary clashes are accepted ingredients of this improvised art.

The second example is from *Laird Baird,* a themeless blues in B$^{\flat}$.[29] The central feature of this excerpt, besides Parker's rhythmically ingenious improvised melody, is the turnaround, a chord progression designed to connect the tonic of measure 11 with the tonic chord at the start of the next chorus. Pianists and bassists have several stock progressions from which to choose. Here Hank Jones and Teddy Kotick agree perfectly on a favorite bebop turnaround: a series of 5th-related enriched dominant sevenths ($G+^{7(\flat 9)}$—C^{9}—$F+^{7(\flat 9)}$). It is Jones's moment; Parker leaves a large space for him to fill, and Roach's snare-drum

fills are quiet and supportive. Once past the turnaround, Jones recedes into the background to play the enriched blues progression that the players follow throughout the performance. Here again are harmonic clashes, at the start and near the end of the example, between Jones's passing chords and Kotick's more basic choices. But again, the clashes over harmonic details do not matter, for the players agree on the main harmonic events.

Charlie Parker, the lead player in both of these excerpts, was a vitally important player during the early years of bebop. Without him the music would have been far different. To know his personal idiom is to know much about the language of bebop. Thus, the following discussion of his musical vocabulary serves as the central point of reference for the rest of the book.

CHAPTER THREE

The Parker Style

Charlie Parker was the architect of the style. (Dizzy Gillespie[1])

Just as Louis Armstrong dominated jazz in the 1920s, so Charlie Parker (1920–55) dominated early bebop in the 1940s. His mastery of the new musical idiom was complete, and his style had a self-contained logic beyond that of any of his contemporaries. Like Armstrong before him, Parker was a principal role model for jazz players worldwide. Elements of his style were copied, not only by innumerable alto saxophonists, but by tenor and baritone saxophonists, clarinetists, trumpeters, pianists, and others. Countless players learned his recorded solos note for note, and sometimes even recorded them, or variations on them.

Parker's musical beginnings in Kansas City foretold little of the greatness he was to attain. He received his meager formal musical training in public schools, where his abilities impressed no one at first. His jazz training came during his teenage years, and primarily consisted of listening to the Count Basie band and other jazz groups appearing in local night clubs, studying jazz recordings, practicing, and playing in some small groups. He progressed quickly, for at age 17, in late 1937, he joined Jay McShann's band. After a few months he quit the band to go to New York City, where he played intermittently. In 1940 he rejoined McShann and toured with him until July 1942. During visits to New York City he participated frequently in the seminal after-hours jam sessions at Monroe's Uptown House and Minton's Playhouse (see pp. 8–10). After leaving McShann, he worked in a variety of bands led by Earl Hines, Noble Sissle, Billy Eckstine, and others. By 1945 he was working mostly with small groups of bebop players. Late in the year he left New York with a group led by Dizzy Gillespie, to play his first Los Angeles engagement. He stayed on the West Coast for a year and a half.

In mid-1947 he returned to New York City to stay, although he often toured various parts of the country, and visited Europe briefly in 1949 and 1950.

Parker's recorded legacy begins in 1940, with two unaccompanied solos recorded on amateur equipment, and some private recordings by the Jay McShann band for broadcast on a Kansas radio station. Other broadcasts and some studio recordings followed in 1941 and 1942, always with the McShann band. These few pieces plus a few from some jam sessions are the only known recordings from his early period.[2] They reveal that Parker was learning his craft by using elements of the musical language of swing. Fascinating as these early recordings are for providing a glimpse into Parker's musical evolution, they had little impact on his colleagues. His influential recordings began in September 1944 with a Savoy recording date under the leadership of guitarist Tiny Grimes, and ended ten years later with a Verve recording of pieces written by Cole Porter.[3] Over 900 samples of his art, ranging in length from brief aborted studio takes to four- or five-minute solos from public performances, date from those ten years. Within this corpus is a wealth of material for study.[4]

Perhaps the first feature of Parker's style to strike the ears is his tone quality; compared with that of his swing-era predecessors, it was harsh, hard-edged. Its stark beauty contrasted sharply with the sweet and mellow timbres used by Johnny Hodges, Benny Carter, and other older alto saxophonists. His vibrato also departed from the swing-era norm; he used a narrower pitch range (about 120 cents) and a slower speed (about five oscillations per second, compared with about six per second for most of his predecessors).[5] Also, unlike his predecessors, he rarely played a note long enough to warm it with vibrato.

Another striking feature is the rhythmic aspect of his solos. In tempos of ♩ = 200 or more, he usually played long strings of swing eighth notes (or even eighth notes if the tempo was fast), articulated in a weak-to-strong manner:[6]

Further, he often accented the highest note of the moment, whether it fell on the beat or between beats, and thereby produced a rich variety of rhythms within those long strings of eighth notes. None of these features was new to jazz when Parker employed them; Armstrong often accented the highest notes of phrases, Sidney Bechet used the weak-to-strong phrasing of eighth notes, and Coleman Hawkins was fond of long strings of swing eighth notes. But the particular combination of these ingredients produced a distinctive sound. In more moderate tempos (♩ = 125–200) he usually intermingled eighth-note and sixteenth-note phrases in varying proportions. In tempos below ♩ = 125 most of his phrases were in sixteenth and thirty-second notes. His skill in playing many notes per second and organizing them into coherent and interesting phrases was extraordinary in the 1940s, for few players could equal him in this regard.

Parker, like all important improvisers, developed a personal repertory of melodic formulas that he used in the course of improvising. He found many ways to reshape, combine, and phrase these formulas, so that no two choruses were just alike. But his "spontaneous" performances were actually precomposed in part. This preparation was absolutely necessary, for no one can create fluent, coherent melodies in real time without having a well-rehearsed bag of melodic tricks ready. His well-practiced melodic patterns are essential identifiers of his style.

Because Parker's recorded legacy is extensive it is possible not only to compile a list of Parker's favorite figures but to see how he used these figures in different contexts. A study of his hundreds of choruses of the blues in B♭, F, and C and dozens of choruses of *I Got Rhythm, What Is This Thing Called Love?, How High the Moon,* and others shows that his improvising was largely formulaic. The specifics of the theme were rarely significant in shaping his solo; instead, he favored a certain repertory of formulas for the blues in B♭, a slightly different repertory for the blues in F, a much different one for *A Night in Tunisia* in D minor, and so on. Some phrases in his vocabulary came from swing, either unchanged or modified; others he created. But whether using borrowed or original melodic formulas, his way of combining and organizing them was his own.

Parker's formulas fall into several categories. Some are only a few notes long and are adaptable to many harmonic contexts. They tend to be the figures he (and his imitators) used most often, for they occur in many different keys and pieces. Others form complete phrases with well-defined harmonic implications, and are correspondingly rare. Most occur on a variety of pitches, but others appear on only one or two pitch levels. A few occur only in a single group of pieces in a single key. Included below are his most common figures plus a few of the rare ones that are particularly idiosyncratic.[7]

Two short figures, 1A and 2A, occur more frequently than any others. Each appears once every eight or nine measures, on the average, in his solos. The first (which Coleman Hawkins also used) is an ascending arpeggio, usually played as a triplet, as in the first two versions. When preceded by an upper or lower neighbor, as shown in the second and third versions, it frequently begins a phrase, although it may occur anywhere in a phrase. Figure 2A, an inverted mordent (shown in three different variants), is even simpler in construction and easier to play than the ascending arpeggio of 1A. Because of its brevity and simplicity the motive appears in almost any context, and is an incidental component in a number of more complex figures, such as 2B. Parker's early role model, Lester Young, also made extensive use of 2A, and even played 2B in his 1936 solo in *Shoe Shine Boy.*[8] This latter figure occurs at the beginning of Parker's solo in *The Jumpin' Blues* of 1942, and reappears as the beginning of the *Ornithology* theme in 1946. Figure 1B contains a rising arpeggio, as in 1A, but turns back on itself. Though it is as easy to play as 1A, it appears less than one-fifth as often.

Ranking third in order of frequency is the more extended Figure 3. It occurs in several different pitch shapes, all of which contain a characteristic feature of the harmonic minor scale: the raised 7th and lowered 6th scale degrees (B and A♭ in this example), which function as the 3rd and minor 9th of a dominant minor-ninth chord. Because of its strong harmonic implications this figure nearly always occurs where there is a V^7-I harmonic relationship, especially a secondary dominant relationship, such as the [V^7-I]ii progression in measures 8 and 9 of the blues (G^7-Cmi in the blues in B$^\flat$).

The chromatic motions of Figures 4A–F are hardly distinctive; pointing them out is rather like pointing out the frequent use of some common prepositions in literary works. For example, the five-note chromatic ascent of Figure 4F was a favorite of Lester Young as well as Parker. Yet these bits are important components of the language; in particular, the simplest of these Figures, 4A, is part of what David Baker has termed the "bebop dominant scale" and the "bebop major scale."[9]

The chromatic Figure 5A appears occasionally in solos by Coleman Hawkins and Roy Eldridge, but often in the rhythmic configuration of quarter-eighth-eighth, not the three eighth notes that Parker favored. Figure 5B is more distinctive and became one of the most frequently copied of Parker's figures. He may have learned it from the beginning of Ellington's *Concerto for Cootie* (or *Do Nothin' Til You Hear from Me*). Figure 5C is a combination of 1A and an extension of 5B.

9

10

11A 11B

Briefly, Figures 6A and 6B are common tonic chord embellishments; Figure 7—a short-note ascent or descent—is a frequent phrase beginning; Figure 8 incorporates both the raised 9th and lowered 9th of a dominant chord; Figure 9 became a bebop cliché for approaching the ii chord in measure 9 of the blues; Figure 10 occurs most frequently over the IV♭7 chord (as in measure 5 of the blues); and Figures 11A and 11B typically occur over the ii-V progression.

12

The famous "flatted fifth" of bebop played a relatively small role in Parker's playing; many solos contain not a single instance of it. But when he did use it he usually called attention to it in a striking way, as shown in Figure 12.

13A 13B

14

15A

15B

16

The remaining figures shown above are less common but aurally striking, nonetheless: 13A and 13B emphasize the augmented 5th of a dominant chord, as in the progression [V^7-I]IV; Figure 14 touches on the 13th, augmented 11th, and 9th of a dominant chord; Figures 15A and 15B are built upon diminished 7th chords (although Parker was probably thinking of dominant sevenths with flatted ninths in 15A); and Figure 16 is built upon a whole-tone scale.

In addition to favoring the patterns shown above, Parker was fond of quoting bits of melodies, often with humorous intent. In doing so he was following a tradition well established in earlier jazz by Louis Armstrong and others. Usually these quotations appear in the informal performances, although some appear in the studio recordings as well. His favorites were the opening of the clarinet descant of *High Society*, the opening phrase of the *Habañera* of Bizet's opera *Carmen*, and the opening of Grainger's *Country Gardens,* which he used over and over again as a coda.

Parker's solos often have an inevitability about them; they seem to have been created by a man who knew exactly where he was going and what to play next. His style seems to have internal logic and consistency. The great number of players who copied his style probably sensed this quality in his music, which is why they copied him in the first place. But what makes his style so consistent within itself?

Much of the compelling nature of his improvising had to do with his command of the instrument and of his musical language. He had the technical equipment to play whatever he wanted to play, whenever he wanted to play it. And he had a sense of swing that almost never wavered.[10] Even when his personal life and physical condition were in chaos his capacity to make coherent music was nearly unshakable.[11] The sheer energy and self-confidence of his playing surely contributed to his music's impact on listeners.

But lying beneath the surface of most of his improvisations is another factor that helps generate the sense of rightness in his music. Typically entire phrases, and even entire choruses and groupings of choruses, are goal-oriented; they arrive on a final note that lies at the end of a lengthy stepwise descent.

To see this scalar descent in microcosm, one need only re-examine some of his favorite melodic formulas. In Figure 3

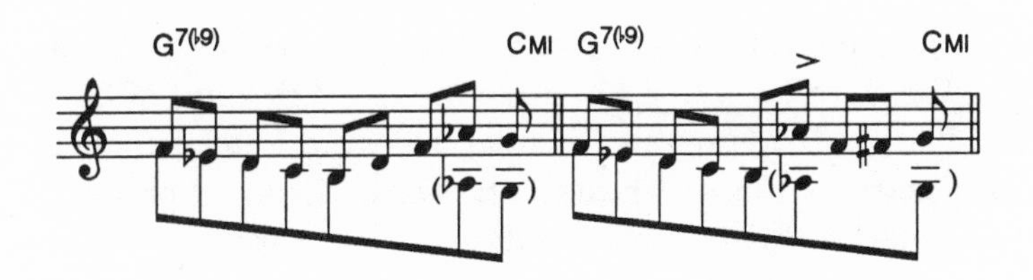

there is a five-note scalar descent to the third of the dominant chord (B in this example), then a rise (sometimes filled in by chordal leaps, sometimes not) to the minor ninth of the dominant chord (A♭ in this example) and a final descent to the fifth of the tonic chord (G here). Thus, there is an underlying descent of a seventh, folded over to stay within the mid-range of the saxophone. In Figures 4A, 4D, and 4E the short descents are obvious, and they are only slightly veiled in Figures 5A, 5B, 5C, 9, and 10. In 11A and 11B they are embedded within broken chord figuration. In 15A the central descent is from E♭ to E; but the initial high F and the final D are also part of this descent, though the octave shifts and chordal leaps hide this connection.

To see fully how these scalar descents work, one must delve into a solo. Here is the opening chorus of his great solo in *The Closer,*

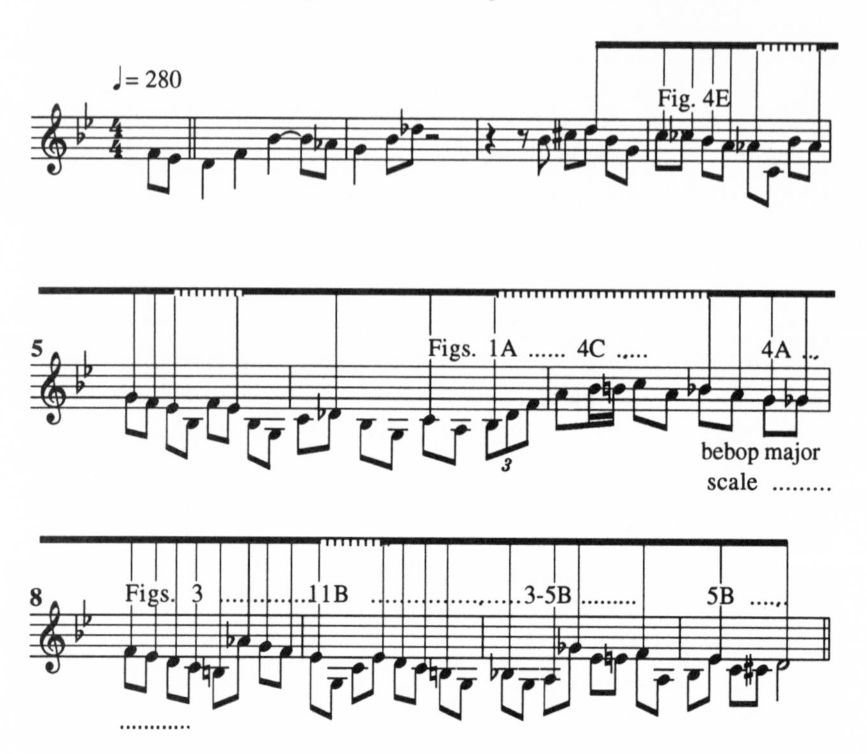

a blues from the Jazz at the Philharmonic concert of 18 September 1949.[12] The downward stems indicate the actual rhythms of his improvised melody; the upward stems connected by an extended beam indicate the notes of the scalar descent, which begins in the third full measure. The breaks in the beam indicate places where he interrupts the scalar descent and then returns to the note on which he left off. He ingeniously disguises the descent with an unpredictable array of chordal leaps, shifts of register, and repetitions. The shifts of register, such as the A to G♭ in measure 10, are dramatically effective. They are also necessary, for unfolded, the descent that underlies this chorus would span four octaves!

Not all of his solos exhibit a scalar organization as clearly as this one. Slow solos, such as the beautiful *Parker's Mood,* take 2 (see below), mix scalar descents with other motions. And some faster pieces have upward as well as downward scalar activity. But the great majority of his solos contain extensive sections of scalar descents; they are among the most striking elements in his musical vocabulary. Further, this scalar organization is a device that he brought into jazz, for his predecessors' music does not contain it.

Was he aware of this aspect of his music? He never spoke of it publicly, and his colleagues have not reported him discussing it with them. But surely he knew, if at a subconscious and nonverbal level, of the rightness of certain choices of notes at certain moments in musical time. And just as it has been with other great makers of music, Parker's choices turn out to have a logic of their own, a logic that extends beyond simply applying melodic formulas at harmonically appropriate times (notice how effortlessly his melody in *The Closer* seems to glide from one pattern to the next, especially in measures 6 through 11). Sensitive listeners—including the many players who incorporated this procedure into their personal styles—may sense that logic intuitively, and analysis reveals at least part of that logic.

The preceding discussion of Parker's techniques of improvisation may suggest that the individual performance was of little importance, that once he arrived at his system of improvising he could produce great solos consistently and almost effortlessly. There is some truth in this supposition; the many alternate takes available for study allows us to hear one brilliant solo after another, created within minutes of one another on the same chord structure. How does one decide which solo is the best among the various takes of *Out of Nowhere, Chi Chi,* or any number of other pieces? But there also were moments of particular brilliance when he surpassed himself. Here is a list of some of his finest recorded moments:

Koko, take 2 (*Cherokee*)—26 November 1945; Savoy 597 (reissued on Savoy CD 70737)

Ornithology, take 4 (*How High the Moon*) and *A Night in Tunisia,* take 5—28 March 1946; Dial 1002 (reissued on Spotlite 101)

A Night in Tunisia, Dizzy Atmosphere (*I Got Rhythm*), and *Groovin' High* (*Whispering*)—29 September 1947; Roost 2234 (and a variety of bootleg reissues)—a broadcast performance from Carnegie Hall

Embraceable You, 2 takes (ballad)—28 October 1947; Dial 1024 (reissued on Spotlite 104)

Out of Nowhere, 3 takes, and *Don't Blame Me* (ballad)—4 November 1947; Dial 1021 and LPs (reissued on Spotlite 105)

Donna Lee (*Indiana*)—8 November 1947; Spotlite 108—a broadcast performance

Parker's Mood, the 2 complete takes (slow blues)—18 September 1948; Savoy 936 and 12000 (reissued on Savoy 5500)

Perhaps, the 4 complete takes (blues)—24 September 1948; Savoy 938 and various LPs (including Savoy 5500)

Salt Peanuts (*I Got Rhythm*)—12 December 1948; Charlie Parker Records 701B and other bootleg records—a broadcast performance from the Royal Roost

Scrapple from the Apple (*a* section from *Honeysuckle Rose, b* section from *I Got Rhythm*)—15 January 1949; Charlie Parker Records 701C and various bootleg reissues—a broadcast performance from the Royal Roost

The Closer (blues)—18 September 1949; Mercury 35013 (reissued on Verve CD set 837 141-2)—a Jazz at the Philharmonic concert performance

Au privave,[13] 2 takes (blues)—17 January 1951; Mercury/Clef 11087 (reissued on Verve CD set 837 141-2)

Anthropology (*I Got Rhythm*)—31 March 1951; Alamac 2430 and other bootleg reissues—a broadcast performance from Birdland

What Is This Thing Called Love?—25 March 1952; Mercury/Clef 11102 (reissued on Verve CD set 837 141-2)—a big-band recording

Hot House (*What Is This Thing Called Love?*) and *A Night in Tunisia*—15 May 1953; Debut 4 (reissued on many bootleg LPs)—concert performances in Massey Hall, Toronto

Chi Chi, 3 takes (blues)—4 August 1953; Clef and Verve issues (reissued on Verve CD set 837 141-2)

On pp. 17–19, I discussed some aspects of the first piece on this list, *Koko.* In the main, *Koko* is Parker's two-chorus improvisation on the chords of Ray Noble's *Cherokee,* which Count Basie's and Charlie Barnet's bands recorded in 1939. This harmonic structure looms large in the literature written about Parker, primarily because of an article that appeared in *Down Beat* in 1949. The authors, Michael Levin and John S. Wilson, wrote:

> Charlie's horn first came alive in a chili house on Seventh Avenue between 139th street and 140th street in December 1939. . . . Working over *Cherokee* with [guitarist Biddy] Fleet, Charlie suddenly found that by using higher intervals of a chord as a melody line and backing them with appropriately related [chord] changes, he could play this thing he had been "hearing." Fleet picked it up behind him and bop was born.[14]

Soon other writers took this third-person description and rewrote it so that it seemed to come from Parker's mouth (see, for example, Shapiro/Hentoff 1955: 354). Since then it has become one of the key "quotations" used to explain the origins of bebop. It is a fuzzy statement, no matter who first made it. "Higher intervals" presumably refers to 9ths, 11ths, and 13ths added to simpler chords. But harmonic enrichment of this sort was part of jazz for years before 1939–in Bix Beiderbecke's sometimes awkward borrowings from the music of the French Impressionists, in Ellington's compositions, in Tatum's ornate runs, and elsewhere. "Backing them with appropriately related [chord] changes" is almost meaningless. The mental image of a saxophonist backing himself with *any* chord changes while playing his saxophone in a pre-electronic-music age is surrealistic. (Perhaps Parker in 1939 had a vision of the MIDI-ied musical environment of the 1980s?) And one wonders how chord changes appropriate to 9ths, 11ths, and 13ths differ from those appropriate to simpler sonorities; both a $G^{13(\flat 9)}$-$C\text{MA}^9$ and a G^7-C are the same basic V-I progression. But the main problem is that those "higher intervals" play a relatively small role in his musical vocabulary. The essence of his style lies elsewhere.

Koko is the most famous of Parker's improvisations on the 64-measure *Cherokee* chord structure. There are two earlier recorded versions, a private recording from the early 1940s and a warm-up piece, issued as *Warming Up a Riff*, from earlier in the *Koko* session. There are also several later performances from radio, concert, and night-club engagements. All have melodic features in common, indicating that during the six years that Parker had been "working over" the *Cherokee* chords he had developed some comfortable ways of moving through those chords. Such preplanning was a necessity, of course. Parker played the piece quite fast (one version is at ♩ = 355), and had to employ some well-practiced melodic ideas in order to avoid stumbling.

The clearest example of a prepared phrase in the 1945 *Koko* occurs at the beginning of his second chorus, where he quotes part of the famous clarinet descant from *High Society* (incidentally, a phrase that simply outlines the tonic triad, with no "higher intervals"). But he does not allow the quote to sit there in bold relief; he integrates it so seamlessly into the fabric of his phrase that anyone unaware of the New Orleans tradition upon which he is drawing will assume the entire phrase is Parker's creation. Another prepared phrase is Figure 3, which he uses in measures 14 and 15 of the first *a* section in each chorus, where the chords are G^7-$C\text{MI}$. A third set of examples occurs four times in measures 3 and 4 of the *a* sections, where the chord is $B^{\flat 7}$, on its way to $E^{\flat}$. He uses Figure 13A, but in three different configurations:

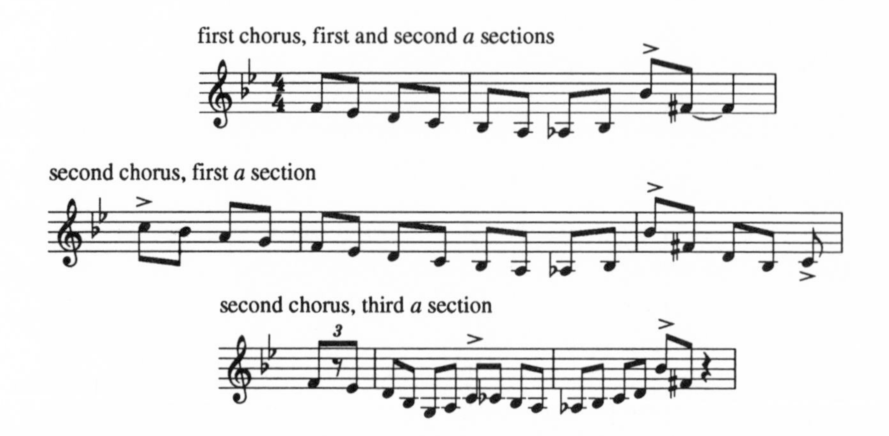

In the first chorus he uses 13A twice to end two nearly identical short phrases. In the second chorus it first appears in the middle of a six-measure phrase, and then as a phrase ending once more but with extra notes interpolated to form a longer melodic unit and to make the high note land on a strong beat.

There are other interesting features here, such as the varied sequential repetitions in the bridges, where he moves rapidly through the ii-V-I progression in the keys of B, A, and G. The thousands of musicians who have learned this solo by ear or played through it from written transcriptions have experienced these features most intensely. The melodic integrity and relentless forward momentum of this solo are awe-inspiring, especially to all who try to play it at the original tempo of ♩ = 300. To those players no further evidence of Parker's greatness is needed.

Several of the recordings in the list above had no historical impact, for they were issued after Parker's death. One could debate the ethics of praising performances that are alternate takes rejected by the performers or that are public performances issued on bootleg records, for the musicians probably received no compensation for these recordings. But some of these pieces give us a more complete picture of Parker's art: 1) they confirm that what he played in the studio was, in the main, how he played in public performance; 2) frequently they provide us with longer solos than are contained in the studio recordings, and thus they show how well he could sustain an extended solo; and 3) in some cases they provide us with gems that we would otherwise never hear.

The supreme affirmation of this last point is *Parker's Mood,* take 1, the first of five takes of a slow blues, in which his accompanists were pianist John Lewis, bassist Curly Russell, and drummer Max Roach. Apparently rejected because of a mistake in the coda—the released version cuts off abruptly at the point where the coda begins on the "master" take—it is complete otherwise and contains the most expressive solos Parker ever recorded.[15]

Take 1 of *Parker's Mood* begins disappointingly with a pompous two-measure phrase, the only precomposed portion of the piece (and therefore, unfortunately, the portion by which the piece is most easily identified). Then the tempo drops abruptly to ♩ = 65, the tempo for the remainder of the piece, and Lewis establishes the proper mood for Parker's solo with four additional introductory measures. Following the introduction Parker improvises two florid choruses filled with wonderfully irregular time values and expressive variations of pitch.[16] Both choruses begin with inventive three-fold, varied repetitions of phrases that immediately command attention. The solo contains a number of his standard melodic figures, but the slow tempo gives him more time than usual to reshape and combine them, and to think of new phrases. In the process he creates a beautiful and poignant picture of the poetic meaning of the blues—he "tells his story" as though he were a great blues singer. His pitch inflections in the last two measures of each chorus are particularly touching.

After these superbly expressive choruses, Lewis's chorus sounds pale and superficial; Parker's bittersweet lament gives way here to an inappropriate lightheartedness. And when Parker returns for the final chorus he is less serious, perhaps because Lewis has altered the mood. This final chorus is more straightforward rhythmically, due in part to Max Roach's double timing (replacing $\frac{4}{4}$ with $\frac{8}{8}$),[17] but is less profound than the first two. At the end Parker uses a two-measure cadential formula that is nearly as old as jazz itself. The take ends abruptly as Lewis hits a GMI triad, the first chord of both the introduction and the coda (although we do not get to hear the coda in this take).

By the time the quartet was able to complete a take, the one ultimately issued as the "master," the tempo had moved up to ♩ = 80, and Parker's mood had lightened. Although he plays well, his solos lack some of the emotional involvement he achieved earlier. Of course the public had no way to compare takes until the longest of the alternates were issued in a "memorial" album upon his death. And the master take became well known; soon beboppers viewed those choruses as community property, and have been drawing upon them ever since. Sometimes the borrowing is literal and extensive, as when singer King Pleasure added lyrics to the complete choruses, or bassist Buddy Clark scored them for the five saxophones of the group Supersax.[18] More often individual phrases are borrowed intact or paraphrased, as when Jackie McLean, Frank Morgan, and countless others pay homage to their musical hero (see the next chapter).

His solos in *Ornithology,* take 4, and *A Night in Tunisia,* takes 1, 4, and 5, have neither the sustained brilliance of *Koko* nor the expressivity of *Parker's Mood;* each is noteworthy mainly because it contains one stunning phrase. The *Ornithology* solo builds gradually toward its climax. It begins with two perfectly balanced phrases, each filling three measures and resting for one, and each having the same pitch contour. Next comes a longer phrase. Then, three measures before the halfway point in the chorus, he begins a phrase with two of his favorite figures (Figs. 9 and 4E), which soon give way to some startlingly complex rhythms:

This rhythmically convoluted phrase spills over well into the second half of the chorus, at measure 17. Unlike most of his vocabulary, it appears to be a once-in-a-lifetime burst of creativity, for I find it nowhere else in his œuvre. The rest of the solo is good, but cannot match the excitement of this spectacular phrase.

Another great phrase occurs in *A Night in Tunisia,* from the same recording session, but its place in his music is far different. Parker's first recording of this piece came a month after Gillespie recorded it with a septet, and he adopted most of Gillespie's arrangement. Both recordings begin with the same Latin-tinged introduction and theme statement, and both abandon the Latin rhythm at the interlude. Next Parker plays a four-measure solo break (Gillespie's were two measures long) that is as stunning as any in recorded jazz. It is an unbroken stream of sixteenth notes—twelve notes per second for five seconds straight—that forms a perfectly structured melodic statement. After completing an initial take, Parker remarked, "I'll never make that break again" (Russell 1973: 212). But the other musicians made so many mistakes elsewhere in the take that Parker rejected it. The players needed four more takes to get a version acceptable for release. However, Parker's solo in take 1 (entitled

Famous Alto Break) and all of take 4 were released eventually. Comparison of the three solos reveals that Parker did indeed "make that break again"; the three breaks are virtually identical performances of what was apparently a precomposed, memorized phrase (this one is from take 5, the master take):

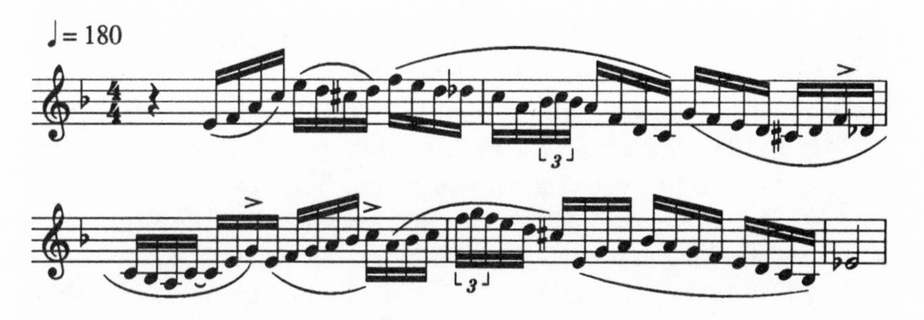

In later concert performances he used variants of this same phrase. It is the longest and most complex non-improvised phrase in all of his solos. Apparently he felt he could not improve upon it. He was right.

Another famous phrase signals the beginning of his most famous ballad recording, *Embraceable You,* take 1. So short and simple that "motive" describes it more accurately than "phrase," it consists of a series of 16th notes, C-G-F-E-D, and a half note, E. As a rule, Parker did not use a single motive as a building block for a solo. But in this piece he gives to the motive an uncommon importance. He states it in various ways five times, on three different pitch levels, in the first five measures before moving on to other ideas.[19] Then he returns to three more variants of it near the beginning of the second half of the chorus. His use of this motive parallels closely the way Gershwin used the D-E-F motive in the original melody. But Gershwin's melody was not Parker's concern here; in this beautiful and ornate fantasia on Gershwin's chords, Parker touches on the original melody only twice, for three beats in measure 21 and two beats in measure 30.

More than half of Parker's recorded legacy is of public performances in night clubs, concert halls, and dance halls, and of private performances in jam sessions and rehearsals. Many of these pieces were recorded with poor equipment, and their acoustical qualities vary from poor to abysmal. Many also are frustratingly incomplete, because the person doing the recording wanted to save tape and was interested only in recording Parker's solos. (The large collection recorded by Dean Benedetti—issued recently on the seven-CD set Mosaic 7-129—illustrates both shortcomings.) But within this hodgepodge are some treasures, which I included in the list above.

One of these treasures is the concert at Carnegie Hall on 29 September 1947 given by Parker, Gillespie, John Lewis, bassist Al McKibbon, and drummer Joe Harris. The pieces are interrupted or incomplete because they ran too long to be captured on one side of a 78-rpm disc. Parker's solos in *A Night in Tunisia, Dizzy Atmosphere,* and *Groovin' High* are at least twice as long as those in the original studio recordings of these pieces. Spurred on by a supportive audience, he creates some wonderful solos. Other great solos originated from the many broadcasts he and his groups did from night clubs, especially the Royal Roost and Birdland (the club named in his honor). For the Birdland broadcast of 31 March 1951 the band was Parker, Gillespie, Bud Powell, bassist Tommy Potter, and drummer Roy Haynes. The recording quality is tolerable and the performances are energetic, polished, and complete.

The best-known of these on-location performances is the concert at Massey Hall in Toronto on 15 May 1953. The players were among the finest in bebop: Parker, Gillespie, Powell, Roach, and bassist Charles Mingus. Overall their best performance is of *Hot House;* the solos by Parker (playing a *plastic* saxophone!), Gillespie, and Powell are superb and Roach's support for them is masterful. *A Night in Tunisia* also contains some wonderful solos. There are some flaws, however: *Salt Peanuts* suffers from a lack of rehearsal[20]; Powell's relentlessly busy and non-swinging chordal backgrounds in *All the Things You Are* seem out of place in this quintet; and the placement of the rhythm section's microphone slights Mingus's bass lines. Mingus tried to remedy this last problem, as he prepared the tape for release, by overdubbing his part on some of the pieces.[21] The overdubbing works well in *Hot House,* providing a level of support for the group that is rare in recordings of the early 1950s. But he has trouble staying synchronized with the recording in the theme of *A Night in Tunisia* and in parts of *Wee.*

Whether in the recording studio, on a night-club stage, or even in an informal rehearsal or jam session, Parker was superbly disciplined while in the act of making music. He was decidedly less successful in other aspects of life, however. He abused his body with alcohol and other drugs off and on throughout his adult life, a life complicated by ulcers, cirrhosis of the liver, debts, broken marriages, suicide attempts, and institutionalizations. To be sure, there were good times, times when he had a handsome income playing in Norman Granz's Jazz at the Philharmonic concerts, times when he was temporarily free of his addictions and could lead a normal domestic life. But the problems took their ultimate toll on 12 March 1955. Sources disagree as to the exact cause of death; it may have been stomach ulcers, pneumonia, advanced cirrhosis, a heart attack, or any combination thereof. He was 34.

Shortly after Parker died, the phrase "Bird lives!" began appearing on walls and subways in New York City (Russell 1973: 363). If anything, that expression, musically speaking, is even more true today than it was in the mid-1950s: 1) more of his recordings are available now than during his lifetime; 2) his musical inventions exist on innumerable recordings by players worldwide who have been copying him for years; and 3) the jazz world now appreciates more fully the contributions he made to the tradition.[22] There will be frequent references to Parker throughout the remainder of this book; there is no other informed way to discuss bebop.

CHAPTER FOUR

Alto Saxophonists

> Yeah, I'm one of Bird's children, absolutely. . . . You've got to remember, I was in New York in 1947; it was impossible to be in that milieu—especially if you were an alto player—and not be touched by Bird. . . . We're all a result of all the people we've ever heard . . . but Bird was the Beethoven of our time. . . .[1]

With these words alto saxophonist Phil Woods reflected on the jazz scene as he had observed it over three decades earlier. Charlie Parker's influence was not limited to alto saxophonists, a point that recurs throughout this book and one that Woods implies. But in this chapter the focus will be on those "children" whose musical resemblances to their "father" are easiest to assess, the alto saxophonists.

Edward "Sonny" Stitt (1924–82) was among the first of these "children." Miles Davis, who heard Stitt in 1943 (that is, before Parker recorded his bebop style), reportedly said that Stitt was already playing in much the same way that he played years later,[2] but the recorded evidence suggests that Stitt copied Parker. Stitt recorded his first solos, on *Oop Bop Sh'Bam* and *That's Earl, Brother*, with Dizzy Gillespie's Sextet on 15 May 1946.[3] By that time Parker already had made many recordings of which the jazz community was well aware. Stitt admitted to being influenced by Parker, in a 1959 interview.[4]

Stitt's 1946 recordings contain very little that Parker had not already played. His favorite melodic formulas were also Parker's, especially Figures 1A, 2A, 3, 4C, 4E, and 5B (pp. 31–32); he applied these formulas to the same harmonic situations that Parker did, and he often connected them in similar ways, resulting in the same scalar descents that Parker used:[5]

Besides the similarities in melodic vocabulary, both men had similar ways of phrasing and articulating notes and similar tone qualities. The resemblances between the two men's styles are such that on at least one LP reissue of *That's Earl, Brother,* Parker is credited as the saxophonist.[6] Whether the record producer was genuinely fooled or was trying for greater sales through deliberate deception, the juxtaposition on the same LP of *That's Earl, Brother* and the May 1945 Parker-Gillespie performance of *Salt Peanuts* probably misled many listeners.

In his earliest recordings Stitt did not copy any of Parker's solos, or even any complete phrases from Parker's solos. Instead, he internalized the components of Parker's vocabulary and used them spontaneously to meet the improvising challenges of each piece. But in his 1949 recording of *Hot House,* Stitt quotes verbatim from Parker's famous recording of 1945. And his 1964 recording of *Koko* on the album *Stitt Plays Bird* shows clearly that he had studied his role model's work—a fact that is hardly surprising, in view of the album's premise.[7]

Stitt began playing tenor and baritone saxophones in 1949–50, probably to escape his image as a Parker follower. His use of the baritone proved to be short-term, but the tenor opened up new lines of musical thought for him, and he played it frequently throughout the remainder of his career. From the beginning his tenor solos contained Parker-like phrases, as would be expected; but he also added other elements, many directly taken from Lester Young. At first in fact, he often sounded much like the Young-inspired Zoot Sims and Stan Getz of the same period, but with a harsher tone quality.[8] Eventually, though, he made his most distinctive contributions to jazz on the tenor sax.

If Stitt's saxophone styles in large part are derivative of Parker and others, does that mean we should dismiss him as a second-rate jazz artist? Was he merely a "remote rival" to Parker, as Ross Russell (1973: 340) describes him? No, for he contributed much of value. During the 1940s and early 1950s he helped establish the norms of the bebop vocabulary; Stitt and other perpetuators underscored the supreme importance of Parker's style. Further, Stitt matured into a first-rate improviser. His solo and phrase-trading with Sonny Rollins on *The Eternal Triangle* are continuously inventive. His 1959 collaboration with the Oscar Peterson Trio is a classic, especially *Au privave,* in which, buoyed by the great rhythm team of Oscar Peterson, Ray Brown, and Ed Thigpen, he floats effortlessly through eleven imaginative and brilliantly executed blues choruses.[9] Here he integrates the Parker-like phrases with his own phrases into a convincing whole. From 1960, when he replaced John Coltrane in the Miles Davis Quintet, comes an interesting sampling of an updated Stitt tenor style in *On Green Dolphin Street.* Later, the *Stitt Plays Bird* album, mentioned above, made its mark. This album, like most of Stitt's best efforts, features him in an almost jam-session format with only a rhythm section for backing. Other good albums in the same vein include two recorded in 1972, *So Doggone Good* and *12!.* Even near the end of his life a top rhythm section could inspire him to play excellent bebop, as in his 1980 recording of *Bye, Bye, Blackbird.*[10]

Another Sonny, William "Sonny" Criss (1927–77), clearly was another of Parker's disciples. But unlike Stitt, whose melodic figures, tone quality and phrasing were all very close to these elements of Parker's style, Criss confined his borrowings to melodic phrases. Criss's tone was sweeter than Parker's, his fast and automatically applied vibrato was more typical of swing than of bebop, and his habits of phrasing and articulating were clearly different from Parker's.

Unlike Stitt, Criss took a long time to master his craft; his early recorded solos are often uncomfortably tension-filled. The tension resulted from several factors, one of which was his unusually rapid and prominent vibrato. Another was his favorite embellishment, a mordent played at the beginning of a beat, as in measure 5 of his 1947 solo on *Backbreaker*[11] (based on the chords of *Oh, Lady Be Good*):

Played in this way the rapid oscillation becomes the center of attention; in contrast, his bebop colleagues were more likely to take Lester Young's approach and play the embellishment later in the beat, downplaying its importance and pulling the listener forward into the next beat. But the chief source of tension in his playing was his disconcerting unconcern for, or perhaps inability to keep track of, the meter. The first phrase shown above ends with a stock figure used by many players. But it is one beat out of sync; he should have played it as shown below, or shifted two beats later, placing the tonic note on a strong beat:

This phrase comes at the beginning of a solo performed in a Los Angeles concert performance in which he followed the more experienced Wardell Gray, Al Killian, and Barney Kessel. Perhaps he was nervous. Unfortunately he gets worse as he moves on; the ending of the next eight-measure segment is similarly out of phase, as are several after that. His entire third chorus is out of sync, for he begins one beat early with a figure that any other player would start on the downbeat, and stays one beat off thereafter. Similar problems plague his fourth chorus. Only in the fifth and final chorus does he seem to hear where the rhythm section is placing the downbeat.

Charlie Parker was famous for appearing to turn the beat around, but he always knew what he was doing and always resolved the rhythmic tension after a phrase or a chorus.[12] Criss, in contrast, is a promising amateur, struggling to master the intricacies of a complex musical language. *Backbreaker* was not an isolated example; his solos in *Groovin' High, Bopera* (also known as *Disorder at the Border*), *Indiana,* and others betray similar meter problems. In *After Hours Bop,* a slow blues, he handles the meter well, but his command of the harmonies appears limited, for he makes a colossal mistake ("clam") with great conviction in the eighth measure of his last chorus.[13]

Eventually, however, he solved his rhythmic and harmonic problems, and developed into an excellent soloist. His series of recordings for Imperial, done in 1956, is the work of a skilled and creative musician; *Sunday* is especially fine. Some of his best music dates from a March 1959 session with trombonist Ola Hansen, and the excellent rhythm section of Wynton Kelly, Bob Cranshaw, and Walter Perkins. Often his expressive powers on ballads and slow blues, in which he employs his luscious tone quality and his expressive use of dynamic and pitch inflections, are second to none among alto players; *Willow Weep for Me* and *Paris Blues* are remarkable, as is *Blues in My Heart.*[14]

Unfortunately, the economic side of his career had more downs than ups; he was a first-rate musician who led a second-rate career. He worked for several leaders for short periods in the 1940s and 1950s—Howard McGhee, Johnny Otis, Billy Eckstine, Gerald Wilson, Stan Kenton, Howard Rumsey, and Buddy Rich, among others. In 1948 he toured with the Jazz at the Philharmonic troupe. Intermittently he also led small groups in both the United States and Europe. But there were slow times when he had to take work outside the music profession to earn a living. Tragically, a promising upswing in his professional life in 1977 coincided with his falling prey to stomach cancer. Unable to cope with the painful death he faced, he took his life with a gun.[15]

Phil Woods (born in 1931) numbers himself among "Bird's children," and indeed he is—now. But although Woods's commentary at the beginning of this chapter implies an influence dating from about 1947, his earliest recordings, which date from 1954, show little Parker influence. His tone was too rich and full to be confused with Parker's and his melodic ideas often seemed closer to those of Paul Desmond (discussed below). The main Parker influence discernible occurs in an occasional sixteenth-note outburst.[16] In general his was a polite style—polished and, compared with Parker's, unemotional. He liked to build lengthy phrases by developing a single figure over several measures. Among his early solos are at least three gems: *Get Happy, Strollin' with Pam,* and *Together We Wail.*[17] The first is a nine-chorus, four-minute exploration of a standard song, reharmonized in the minor mode. The second is a seven-chorus blues improvisation, and the third shows Woods's great facility and creativity at fast tempos.

By 1957, the Parker influences emerges strongly in Woods's music. It is extensive and clear in a night-club concert by Cecil Payne's sextet devoted to Parker's music.[18] When Woods plays the famous introduction to *Parker's Mood,* he recreates the tone and articulations of his model with remarkable accuracy. In the three-chorus solo that follows he continues to apply Parker's tone and phrasing habits to a largely fresh melodic vocabulary. The effect is almost that of Parker himself playing yet another slow blues solo. (Remember that the various takes that Parker made of that slow blues contain no fixed theme.) Woods also plays the other pieces included on the recording in much the same way that Parker played them—with nearly the same tone quality and articulation habits and many of Parker's favorite figures.

Later recordings confirm that Woods's mature style was to be Parker-derived but not slavishly so. Often a phrase borrowed from Parker leads smoothly into something different, as in this example from *Now's the Time:*[19]

Here the first two measures are pure Parker, but the rest is an unParker-like development of a three-note motive. And in *Doxy,* take 1, he develops a Parker figure during six measures (Parker would usually play a figure once and then move on to something else) before ending the phrase with a Parker-like scalar descent:[20]

In *Altology* he borrows the opening of *Habañera* from Bizet's *Carmen*, which Parker also borrowed often; then a chorus later he plays a variation on it that Parker never imagined. A striking example of Woods rethinking Parker's vocabulary and procedures occurs in *Pairing Off*, from one of several albums he made with fellow alto saxophonist Gene Quill.[21] Inspired perhaps by a figure near the end of Parker's famous *Koko* solo—

—Woods devotes the first nine measures of his solo to developing this figure:

His second chorus begins with an altered version of the same phrase. More explorations of the same phrase end chorus 3 and reappear intermittently in the next two choruses. This five-chorus solo ends as it began, with a phrase spun from the opening Parker-like figure. There is no Parker solo held together by a unifying figure as this one is by Woods; he presents here a skillful continuation and renewal of Parker's musical language.

In time Woods's musical dialect departed more from Parker's, especially in the area of tone quality. In the late 1960s and early 1970s he often growled while playing, producing a "dirty" tone quality that Parker did not use. He also occasionally used slap tonguing to produce the percussive, popping effect that Coleman Hawkins and others used in the early 1920s. And in his recent recording of *My Man Benny* he paid homage to the senior member of the alto saxophonists' community, Benny Carter, by borrowing some of Carter's phrasing habits.[22]

While there are musical similarities between Parker and Woods, the careers of the two men took opposite paths. During the 1950s and 1960s Woods was one of New York City's busiest recording session saxophonists. His strong sound, accurate intonation, fine reading ability, and first-class soloing style made him an ideal sideman for record dates. During those years he recorded with Quincy Jones, Dizzy Gillespie, Michel Legrand, George Russell, Buddy Rich, Benny Carter, Thelonious Monk, Jimmy Smith, Benny Goodman, Gary Burton, Oliver Nelson, Clark Terry, Oscar Peterson, Sonny Rollins, and many others. One of his finest moments among these many sessions was his beautiful solo on Quincy Jones's *Quintessence.*[23]

All of this sideman activity had a negative side for a player with Woods's strong musical personality, however; during the decade 1958–67 Woods played only four record dates as a leader. Feeling the need for a change of musical scene, he moved to Europe in 1968, and began a period of musical exploration. Soon he formed a quartet, the European Rhythm Machine. With it he played some jazz standards (his solo on *Stolen Moments* is superb) and some "funky" jazz (*The Day When the World . . .*), tried the Varitone on some harmonically chromatic, melodically angular pieces à la Miles Davis's mid-1960s pieces (*Chromatic Banana*), and even delved into the action-jazz idioms of late Coltrane and Ornette Coleman (parts of *And When We Are Young* and other pieces). The artistic high point for this ensemble was the Frankfurt Jazz Festival of 21 March 1970. With tremendous vitality and great skill the four virtuosos—Woods, keyboardist Gordon Beck, bassist Henri Texier, and drummer Daniel Humair—build one unbroken performance out of *Freedom Jazz Dance, Ode to Jean-Louis,* and *Joshua.* They push bebop to the stylistic breaking point (freebop?); and after an intense 40 minutes they return for a fine, funky performance of *The Meeting.* It is a remarkable and important recording.[24]

In 1972 Woods returned to the U.S., but not to the session-playing rat race of his pre-European years. In 1974 he formed a new quartet and returned to traditional bebop. He and his small groups have given us some of the finest jazz of the past twenty years. To hear the inspired interactions of soloists and rhythm section in such pieces as *Cheek to Cheek* and *Changing Partners,* and on the albums *Phil Woods Quartet Volume One, Phil Woods Quartet Live from New York Village Vanguard,* and *Gratitude* is to hear bebop at its best.[25] And hearing his group in person, filling a jazz club with sound without using a public address system, confirms what the recordings document.

Few players have had a better start on the road to becoming a jazz musician than John Lenwood "Jackie" McLean (born in 1932). His father was a big-band musician; his stepfather owned a record store; he was given soprano and alto saxophones when he was about four; he grew up in Harlem and then on Sugar Hill, where his neighborhood housed

Ellington, Andy Kirk, Nat Cole, Sonny Rollins, and others; Bud Powell befriended and tutored him at age 16.

He began his professional career at age 19 with Miles Davis, with whom he made his first records. They reveal a player well versed in the language of bebop and the dialect of Charlie Parker. However, his tone quality differed from Parker's; also, he had a repertory of unique phrases in his vocabulary, and a very personal concept of intonation. Additionally, he played behind the beat much more consistently than Parker did and favored a detached, almost percussive articulation of notes. Perhaps his finest solo of this early period is on *Dr. Jackle.*[26]

One of McLean's most distinctive musical traits in the mid-1950s was that of playing extremely disjointed phrases over a period of several measures, as in *Sweet Blanche:*[27]

(Notice that he incorporates the beginning of Stephen Foster's *Jeanie with the Light Brown Hair* into this phrase.)

As he matured he gradually divorced himself almost entirely from his early role model. In *Sister Rebecca,*[28] a blues in D minor, the detached phrasing of eighth notes, the tone quality, and the melodic ideas are all different from Parker's habits. In some pieces scattered over a series of albums made under his leadership for Blue Note Records during the 1960s, he moved further away from mainstream bebop, often utilizing a highly chromatic or rhythmically complex vocabulary. For example, in *Cancellation,* he devotes most of his solo to "outside" playing[29] and a remarkably flexible approach to intonation. Evidence that he was listening to John Coltrane's 1960s music is strong in such pieces as *'Snuff,* where he uses Coltrane's long-note phrasing opening (Fig. 13, p. 94) and Coltrane's approach to motivic development. In some pieces he moves into the post-bebop realm of action jazz, as on his album with Ornette Coleman playing trumpet. Yet other pieces in the same series of albums are completely within the bebop tradition of the 1950s.[30]

Over the years McLean has synthesized elements of Parker, Coltrane, Coleman, and others with his own ideas and has developed an energetic, vital style that, for all its flirtations with action jazz, is within the bebop frame of reference most of the time. Once he would have objected to this assessment; in 1967 he said, "I don't want to hear any more about bebop or hard bop or this or that category. Titles hang things up. The music is just good or bad."[31] But recently he seems to have accepted the label: "Be-bop is a language and that's the way I approach it at the university level."[32] He now speaks of the tradition he has learned from Parker, Coltrane, and others, and since the early 1970s he has dedicated himself to teaching that tradition at the University of Hartford. And surely anyone who hears him plumb the depths of the language when he solos on *Lover, Parker's Mood,* or *Confirmation* hears a bebop master of the first rank.

In 1955, just three months after Parker died, a combination of unintentional timing and some recording publicists eager to find a lucrative gimmick made Julian "Cannonball" Adderley (1928–75) appear to be one of Bird's most devoted "children." When Adderley arrived in New York City he was an almost complete unknown whose previous jazz experiences had been in Army groups and local bands in Florida. On the first night of what was to be a summer of graduate study at New York University, he sat in at the Cafe Bohemia with Jimmy Cleveland, Horace Silver, Oscar Pettiford, and Kenny Clarke.[33] His playing impressed the players and audience as sensational; he was offered a job with the group and accepted on the spot, discarding his plans for graduate study. Before the year ended he signed a recording contract,[34] recorded six albums (including three in the capacity of leader), and formed a touring quintet with his brother, cornetist Nat Adderley. EmArcy Records promoted him as "The New Bird," much to his dismay and embarrassment.

Those early recordings show why the initial reaction to his performance at the Cafe Bohemia that night in June was so strong; this 26-year-old saxophonist played with great agility, precise intonation, a rich, warm tone quality, and a fully formed style. His first album as leader contains two excellent blues solos, on *Spontaneous Combustion* and *Still Talking' to Ya.*[35] They show that Adderley surely had learned much from Parker, including phrases employing Figures 1A, 2A, 4A, 4C–F, 5A, 5B, 11A, and 11B (pp. 31–33). Still, he sounded less like Parker than Stitt and others did at the time. From the beginning his tone quality, many of his phrases, and especially his fresh approach to phrasing and articulation set him apart, not only from Parker but from the other alto saxophonists of bebop.

In articulating notes Adderley sometimes divided the beats by using the jerkier 3:1 ratio (♩. ♪), while Parker preferred the smoother 2:1, 3:2, and 1:1 ratios (♩ ♪ (3), ♩. ♩ (5), and ♩ ♩). Also, Adderley often played the notes on the last part of beats staccato, while Parker usually held those notes longer, often slurring them into the next note. The example below shows Parker's and Adderley's different articulations of the first phrase of *Au privave*:[36]

In the next example the shape of the phrase, consisting of a rising arpeggio followed by stepwise descent, is similar to many phrases played by Parker; but the profusion of detached notes is not:[37]

Adderley plays the whole phrase fractionally behind the beat (indicated by the grace note and dotted line), which was a tension-producing device common to Dexter Gordon, John Coltrane, and others in the 1950s. In fact, the combination of separately articulated notes played behind the beat, with upward lip slurs on some notes, appears in Adderley and Gordon solos as early as 1955.[38]

Adderley's greatest period of recorded creativity was approximately 1958–63. This peak of creative accomplishment followed a bleak period during which the Adderleys had struggled unsuccessfully to make a living with their quintet. In the fall of 1957 they disbanded; Julian joined the great Miles Davis Sextet while Nat joined J. J. Johnson. On 4 and 6 March of the following year the brothers reunited for their final EmArcy recordings together, and produced their best album for that company. It contains excellent solos by Julian on *Our Delight, Straight No Chaser,* and the lengthier version of the blues *Fuller Bop Man.* A few days later, with his new boss Miles Davis as his front-line partner, he recorded fine solos on *Autumn Leaves, One for Daddy-O,* and especially *Love for Sale.* The following month he began the first of a series of recordings with Davis's sextet; two of his best from the first date are *Two Bass Hit* and *Miles,* the latter better known by its incorrect title, *Milestones.* Some sessions under his name with first-class ad hoc groups produced *A Little Taste* and a dazzling *Groovin' High.*[39] Early the following year he made invaluable contributions to the classic recordings issued under the title *Kind of Blue* (see below).

After two years, Adderley quit Davis's band to try again as a leader. This time he and his brother were much more successful. A few weeks after organizing their quintet they recorded an album in concert at the Jazz Workshop in San Francisco. The powerful rhythm section of pianist Bobby Timmons, bassist Sam Jones, and drummer Louis Hayes inspired Adderley to do his best, especially in *Spontaneous Combustion* and *This Here.*[40] The latter piece, a 16-measure blues in triple meter, with obvious borrowings from black gospel music, became a hit within the jazz world and boosted the fortunes of the group considerably.

Two studio sessions after the first of the year produced *Dat Dere, Jeannine, Work Song,* and *Them Dirty Blues.* Nat Adderley's folk-like tune *Work Song* became another hit for the group, but Julian's best solo of the album was on *Them Dirty Blues,* a slow blues in B♭, close in tempo, mood, and melodic style to *Parker's Mood.* That fall the quintet (with Victor Feldman replacing Timmons) made a second night-club recording, at the Lighthouse in Hermosa Beach, southwest of Los Angeles. A superb sequel to the Jazz Workshop album, it includes *Sack o' Woe, Big "P,"* and *Azule Serape.* And yet another performance recording took place in the following month, but remained unissued until 1984. It also contains great performances of *Big "P", Azule Serape,* and *The Chant.* This quintet played with a drive, enthusiasm, and skill unsurpassed by any bebop group of the time, and these concert recordings capture it at its best.[41]

Early in 1962 the quintet became a sextet, with the addition of woodwind player Yusef Lateef; at about the same time pianist Joe Zawinul joined the group for what was to become a nine-year stay. Adderley's finest work during the year includes three different concert recordings of the triple-meter blues *Gemini* and an energetic version of *Work Song.*[42]

It is difficult to select for discussion only two or three pieces from this wonderful array of Adderley's inspired performances. But the choices made by Kernfeld (1981: I, 74ff.) are excellent; he offers some perceptive commentary on Adderley's solos from the Davis Sextet period, especially regarding *All Blues,* and *Flamenco Sketches* from the monumentally important *Kind of Blue* album.[43]

Adderley's solo in *All Blues,* a folk-like blues in G, represents a combination of motivic and formulaic improvising. The rhythm of his opening motive, and its span of a third—

—provide the raw material for motivic development throughout the four-chorus blues solo. A few of his transformations of this motive are these:

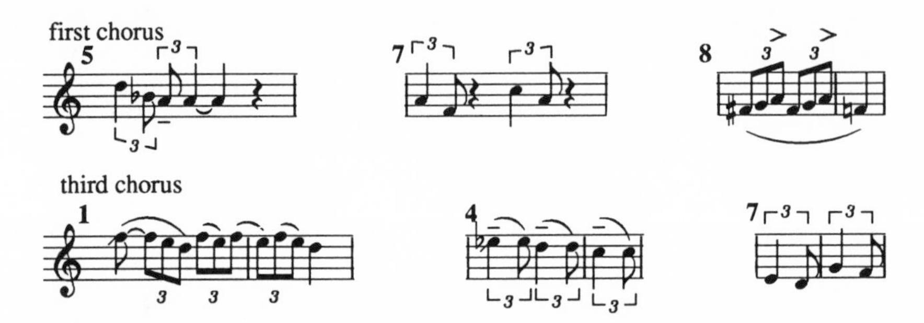

Meanwhile, he punctuates his solo with nine varied statements of this figure:

It is a combination of Figure 11B (see Chapter 3) and cadential figure (the ascending sixth formed by scales steps 3 and 8—B and G here) found in slow blues recordings for decades. Parker and Adderley were only two of many who used it regularly. The solo is an important part of this classic performance, which also contains fine solos by Davis, Coltrane, and Bill Evans.

Kernfeld believes Adderley's solo in *Flamenco Sketches* "is the finest solo by any member of the 1958–59 Davis sextet on any of the sextet albums" (1981: I, 96). This is high praise, indeed, considering the amount of great music that Davis's historic group recorded. He bases his judgment not on a motivically unified structure; there is none, although Adderley does work to some extent with the first three-note figure and one or two others, and skillfully uses scale degree 5 as a melodic center of gravity in the middle segments of the solo. Instead, Kernfeld finds most impressive Adderley's touchingly beautiful lyricism and the subtle, extraordinarily high level of cooperation among Adderley and his accompanists, Bill Evans, Paul Chambers, and Jimmy Cobb. The solo is clearly one of the most magnificent gentle moments in recorded jazz, and is a vital part of a classic ensemble performance (see pp. 225–26 for more on this jazz classic).

Adderley was best known as a funky jazz or soul jazz player, a reputation gained through the playing of a few folk-influenced pieces (such as *This Here* and *Work Song*) and the written publicity gained from them. But during the first years of his recording career these pieces formed only a small part of his repertory. Further, once he finished playing a "funky" theme, with its simple melody based on the minor-pentatonic or blues scale[44] and its harmonies based on triads and dominant seventh chords, and dug into his solo, he and his band almost invariably transformed the piece with the more sophisticated chords and rhythms of bebop. At first he resented being pigeonholed as a soul jazz player. But in the late 1960s the percent of folk-style pieces in his group's repertory increased and some of them were frankly commercial with little or no soloing: Joe Zawinul's compositions *Mercy, Mercy, Mercy* and *Walk Tall* are prime examples.[45] At the same time he explored other musical directions. He began playing the soprano sax and the Varitone electronic attachment. Some moments of action jazz usually surfaced in concert appearances and on records. He recorded some pretentious music for jazz quintet and orchestra. He even sang—unfortunately. But fundamentally he remained a bebopper until his fatal heart attack in 1975. The body of recordings discussed above speaks eloquently; Cannonball Adderley was one of the finest beboppers. Bird surely would have been proud of this musical offspring.

Frank Morgan (born in 1933) made a few records in Los Angeles during 1953–55—with Teddy Charles, Kenny Clarke, and, most important, his own groups.[46] These recordings show that he was another Parker disciple, but one with a distinctive tone quality. Also, he favored separate articulation of each eighth note in moderate tempos (the Dexter Gordon influence?). And in ballads his vibrato was radically different from Parker's; its pitch span was much wider, almost distractingly so.

These early recordings, while showing a few evidences of technical uncertainties, showed a young saxophonist with great promise. What they did not show, however, is that the influence of Parker on Morgan was more than just musical; he also shared Parker's drug problem.[47]

Arrested for the first time in 1953, he spent the next two and a half decades in and out of prisons (including the notorious San Quentin, where he played with fellow saxophonist Art Pepper in the prison band) and drug rehabilitation programs. Thus, there is a gap of 24 years between recording dates, and a gap of 30 years between his first and second albums as leader.

But the recent recordings show that he did not waste the intervening years, for he now plays with great authority, imagination, and control. His soprano playing is impressive on *Fantasy Without Limits* and *Frenzy,* in a 1979 album with L. Subramaniam (Trend 524). And his 1985 album (Contemporary 14013) is full of delights; among them are an effective and at times moving rendition of the title song, *Easy Living,* a marvelous exploitation of pitch inflection and dynamics on *Manhã de Carnaval,* and a lyrical, gentle presentation of McCoy Tyner's little gem, *Three Flowers.* Throughout the album the fine rhythm section of pianist Cedar Walton, bassist Tony Dumas, and drummer Billy Higgins propels the music forward masterfully. The one lapse of taste is his deliberate bow to his early role model; at the end of an otherwise excellent performance of *Embraceable You* he shatters the mood, just as Parker often did, with the trite *Country Gardens* coda.

In the late 1980s Morgan created a series of musically successful albums, including a duet album with pianist George Cables, a concert performance at the Village Vanguard, and a tribute album to Parker (Contemporary 14035, 14026, and 14045). The recordings, however, do not convey fully the beauty of his covered tone played *pianissimo* and the drama of the occasional high note suddenly popping out from under that cover. His dynamic range within single phrases often runs from inaudible to almost piercing, and as he plays he moves about freely, with little concern for microphone location; he must pose a difficult challenge for recording engineers.

As might be expected of a man who has been playing the alto sax for nearly 50 years, he has developed a unique way of using the bebop language. For example, in his wonderful Village Vanguard performance of *Parker's Mood,* he uses a few of Parker's ideas from the master take, but creates a largely independent blues statement.[48] The links to Parker are still there; but just as Adderley, Criss, Woods, and other great musical descendants of Parker did earlier, Morgan has transformed those Parker figures into something uniquely his own, enveloping them in ornate, rhythmically flexible "sheets of sound" that derive neither from Parker nor from anyone else. He has transformed his life into something unique, as well; no other major player has all but destroyed himself and his career and then—30 years later—turned his life around and forged a new and impressive career.[49]

Most of the alto saxophonists discussed so far developed unique musical personalities, but began their recorded careers with strong ties to Charlie Parker's style. A few alto sax players, however, either shed most of their ties to Parker before recording or set out from the beginning with a different stylistic reference in mind.

One of these players was Sahib Shihab (1925–90). His playing with Thelonious Monk in 1947 has much of the sound of a swing-era player, due to his fast and automatic vibrato and his clear, sweet tone quality. A few Parkerisms emerged in his well-polished bebop in the 1950s, but mostly he was an independent. One of his best solos is in the Tadd Dameron recording of *Bula-Beige.*[50] Had he stayed on alto he doubtless would be one of the most respected bebop alto players, but he began playing the baritone sax in 1953, and spent the bulk of his career working as a section player in European big bands and as a copyist.

Gigi Gryce (born in 1927) divided his time between composition—he studied with Daniel Pinkham, Alan Hovhaness, Nadia Boulanger, and Arthur Honegger, and wrote some extended works in European concert style before turning to jazz in the early 1950s—and jazz performance. His tone on alto was thin and subdued, and unlike Parker he favored separate articulations of straight eighth notes. Although in blues and other medium- and fast-tempo pieces he used some of Parker's figures, in slow pieces he betrayed barely a trace of Parker, leaning instead toward the type of sound Art Pepper (see below) was using at the time. Among his best solos are those from his 1955 session with Thelonious Monk; *Nica's Tempo* (Signal 1201) is particularly fine. In the 1960s he left jazz and began a new career as a teacher.

John Handy (born in 1933), while also a tenor saxophonist and saxello player, is primarily an alto saxophonist, one who shed his Parker connections early in his recording career. Like Jackie McLean before him, he first came to the attention of the international jazz audience in the Charles Mingus ensemble. To some extent indebted to Parker, he clearly had a different style by 1959, when he first recorded with Mingus; his performance of *I Can't Get Started* is an excellent independent approach to this often-played ballad. It contains ear-catching features that reappear from time to time in later Handy recordings: a few notes in the altissimo register and a passage played with the machine-gun effect produced by flutter tonguing. In a superb, hard-driving solo in *Blues for M.F.*, he uses a splattered attack that sounds uncannily like a trumpet attack.[51]

At his best Handy is capable of an exquisite lyricism, as in *Dance to the Lady.* He reached a creative high point during his association in the 1970s with sarodist Ali Akbar Khan, playing a modified North Indian style of music in a quartet in which tambura and tabla replaced bass and drums. His elegant sound shines forth on their touchingly beautiful *Karuna Supreme.* While he was not playing bebop in this quartet, his musical roots sometimes came through nonetheless, as in *The Soul and the Atma.* Strangely, during that same period he also delved into fusion jazz (a blend of rock and jazz), and reached a low in creativity but a high in commercial success with a vapid pentatonic riff called *Hard Work.*[52] But in the 1980s he returned to mainstream bebop in a San Francisco sextet called Bebop and Beyond.

The three most prominent non-members of the Parker school were Art Pepper, Lee Konitz, and Paul Desmond. The eldest of these three was Los Angeles-based Pepper (1925–82). Like Parker before him, Pepper's musical point of departure was Lester Young, from whom he apparently got his sweet tone quality, his gentle, even timid approach to high notes, many of his articulation habits, his slow vibrato, and several favorite melodic figures. He was not a slavish imitator of Young, however. He used far more notes per measure than Young, and rhythmically his phrases fit the bebop idiom better than did Young's. In many ways his style in the late 1940s closely resembled Stan Getz's (see Chapter 5). Only one and a half years older than Getz, Pepper recorded his first solo while a member of the Stan Kenton band in 1943, five months before Getz began recording commercially. But he did not start recording many solos until late 1947, when he rejoined Stan Kenton's band after completing his military service. By then Getz had recorded with several well-known groups, and had made his first quartet recordings as a leader. Thus, if either influenced the other it was more likely that Getz influenced Pepper.[53]

During the years 1947–51 Pepper was one of Stan Kenton's main soloists, although his solos were mostly short.[54] And because the Kenton band's popularity then was high in the jazz world, Pepper developed a large following. He placed a strong second in *Down Beat*'s Readers' Poll in 1951, coming in just 14 votes behind Charlie Parker. But it came time to move on; his reasons for leaving are common to many soloists working within big bands: first, he missed his wife, and second,

> I knew all the arrangements by memory and it was really boring. I didn't get a chance to stretch out and play the solos I wanted to play or the tunes. I kept thinking how nice it would be to play with just a rhythm section in a jazz club where I could be the whole thing and do all the creating myself. (Pepper and Pepper 1979: 115)

There are some amateur recordings, made shortly after he left Kenton, that capture him playing in exactly the setting he describes.[55]

The acoustical quality of these recordings is dismal, but Pepper's musicality emerges clearly. The connection with Parker is minimal, even in pieces that Parker regularly played. But the debt to Young is obvious in the moderately slow pieces: the timidly played high notes, the slow vibrato on long notes,[56] and some of Young's pet figures, including the "*Shoe Shine Boy* figure" that Parker also used (Fig. 2B, p. 31). In the faster pieces he plays with more aggressiveness than Young used—more notes, stronger accents, more harmonically derived melodic lines. The quartet albums are particularly good; perhaps inspired by Hampton Hawes's excellent playing, Pepper produces a series of first-class solos. If there is a weakness here it is an occasionally trite phrase ending, which creates the impression that he has run out of ideas.

While the musical connection to Parker is slight, the chemical connection is not. Pepper's drug-related problems, discussed in great detail in his autobiography, kept him on a grim emotional and physical roller-coaster ride. One indication of the turmoil in his life is implied in his discography,[57] which shows prolific and barren periods of recording activity alternating. During the barren periods he was in jail or prison (including San Quentin in the early and mid-1960s), or was so unreliable that producers and leaders were afraid to use him. His final years began with a partial rehabilitation at Synanon, a happy marriage, and a measure of professional stability on the one hand, but a series of health problems triggered by his addictions on the other.

One of those productive periods was July 1956 through January 1958, when he performed on 29 record dates. There are several good recordings from that year and a half, including a relaxed session with tenor saxophonist Warne Marsh, pianist Ronnie Ball, bassist Ben Tucker, and drummer Gary Frommer.[58] But the high point is the album he did with Miles Davis's rhythm section of 1957: pianist Red Garland, bassist Paul Chambers, and drummer "Philly" Joe Jones.[59] In his book Pepper claims that he met the rhythm section for the first time on the day of the recording, that he went into the session with no pieces planned, that he recorded several pieces he had never done before, and that he did the session after not having played his alto for six months (Pepper and Pepper, 1979: 191ff.). The first three facts are probably true; the last is not, for he had played 18 record dates during the previous six months, including one just five days earlier. Nevertheless, there is a sense of others being the actual leaders on some tunes; Red Garland seems in command on *Imagination,* Jones takes over on *Jazz Me Blues* and on *Tin Tin Deo.* This fact does not negate the value of the music, however, for the four men seem compatible, and play some fine bebop.

In some respects his finest album of the 1950s was *Art Pepper +*

Eleven.[60] Recorded during three sessions in 1959, it is a semi-big-band showcase for Pepper, created by the gifted arranger Marty Paich and a polished group of top Los Angeles studio players. The tunes, drawn from the cream of the bebop repertory, include *Airegin, Anthropology, Donna Lee, Four Brothers, 'Round Midnight,* and seven other classics. Pepper's solos on alto and tenor saxophones and clarinet are excellent. Perhaps the high percentage of Parker-related tunes inspired him to think about Parker, for his vocabulary contains a light sprinkling of Figures 1B, 2A, 3, 4D, and 7 (pp. 31–32). Paich's arrangements, while paying homage to the original versions, added effective new colors to these pieces. For *Groovin' High,* he scored Parker's 1945 solo for the sax section, and thus provided an early sample of the sound that Supersax was to explore in the 1970s and 1980s.[61]

Early in 1960 Pepper and the Davis rhythm section held a rematch. By now Wynton Kelly had replaced Garland and Jimmy Cobb had replaced Jones; in addition, trumpeter Conti Candoli played on three of the pieces.[62] This time Pepper was prepared, for he clearly is in command. Among his several superb solos the one on the blues *Whims of Chambers* is the best. It contains few of the gentle gestures of the early 1950s; instead there is a restless quality produced by seemingly disjointed phrases. But closer inspection of this solo reveals that the entire seven choruses contain the development of a single motive—an often-used figure from the end of Paul Chambers's serviceable theme:

The essence of this phrase, to Pepper, is the three-note ascent inside the box. He manipulates these three notes in different ways:

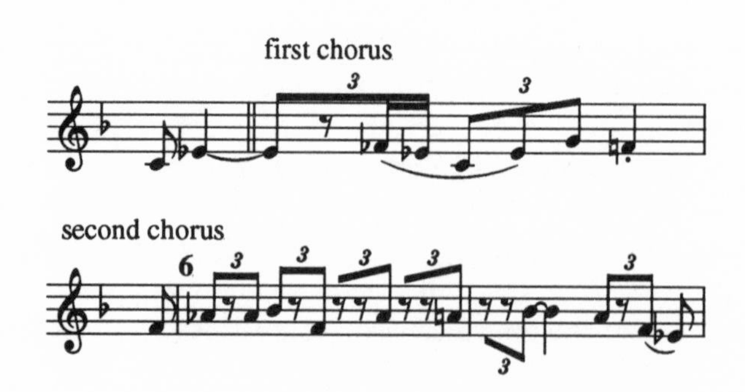

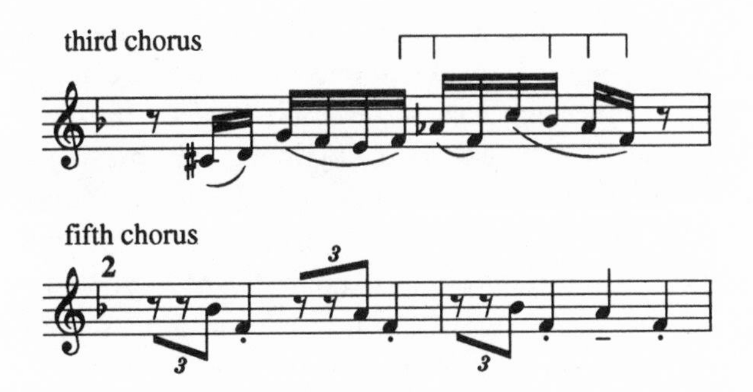

These short excerpts tell only part of the story; over half of his solo, including all of the first and last choruses, contains some form of the motive.

While the motivic unification of this solo is unusual,[63] the employment of short, choppy phrases is not. It had become a characteristic of his style by this time. And more often than not these seemingly disjointed phrases fit together into a logical design. The first miniphrase might consist of two or three notes, the next phrase might repeat those notes with a few notes added, the third phrase might grow out of the second, and so on. His procedure resembles one that Duke Elllington had followed in presenting the lyrics of *The Blues* from the large tone poem *Black, Brown and Beige* in 1943 ("The Blues . . . The Blues ain't . . . The Blues ain't nothing but a cold gray day").

Another change occurred during the late 1950s; his tone developed a harder edge, the timidly played high notes became scarce, and the sweetness turned bittersweet. In view of his chaotic, drug-filled life the change is hardly surprising, although he might have gone through the same musical changes regardless of events in his personal life.

By the mid-1970s his tone quality changed again; it became harsh and intense. His once-fluid melodic lines became jagged and sometimes halting, and his once near-perfect intonation became less reliable. Still, he often played compelling and inventive jazz. His soul-drenched solo in the blues *For Freddie* is a musical high point in the last decade of his life.[64] Using a much faster vibrato, a powerful tone, and large repertory of "funky" licks, he fashioned a compelling solo.[65]

The alto saxophonist least likely to be labeled a "Bird child" is Lee Konitz (born in 1927). He consciously avoided using Parker's vocabulary because in the late 1940s there were already many Parker imitators. Also, he felt that Parker was putting together well-rehearsed figures "like a jigsaw puzzle," and he wished to keep his musical vocabulary more flexible and open to new invention.[66] In fact, however, he had his favorite phrases, and some closely resembled those that Parker used, although his radically different tone, rhythms, and phrasing tended to obscure the connection.

Konitz recorded his first solos with the Claude Thornhill band in 1947. His short solos in *Anthropology* and *Yardbird Suite* reveal a player completely untouched by Parker's idiom, who uses instead a thin tone and almost no syncopation or pitch inflection to fashion his alternately bland and (by Parker standards) bizarre melodies:[67]

Later recordings with Lennie Tristano, the Miles Davis Nonet, and under his own name confirm the initial impression; although he used an occasional figure that Parker used (shown by the dotted lines)[68]—

—the small tone, smooth phrasing, scarcity and gentleness of accents, and mix of un-Parker-like figures created a different aural effect. His restrained, seemingly clinical style epitomized "cool jazz" for many.

If his way with medium and fast pieces was emotionally cool in the early years of his career, his way with ballads was ice-cold. *Rebecca* and *You Go to My Head,* both nearly devoid of rhythmic complexity and pitch inflection and played with his characteristic thin tone and slow, narrow vibrato, are sad and lonely pieces.[69]

But by the mid-1950s his style had undergone some changes. The recordings he made early in 1953 with Gerry Mulligan reveal a Konitz with some rhythmic punch to his playing. His animated five choruses on *Too Marvelous for Words* swing lightly, and he enlivens *Lover Man* with an unusually (for him) wide repertory of rhythms, dynamics, and pitch inflections.[70] In his second chorus in the former piece he plays a three-beat figure that he sequences:

It is only one of several three-beat patterns that he used in three-against-four cross rhythms.[71] Two years later, in the blues *Don't Squawk* and in *Ronnie's Line,* he plays a few blue notes and even uses Figure 5B (see p. 32).[72] Gone were the arbitrarily disjunct melodic lines; smoother lines replaced them. By this time also he was using regularly a terminal vibrato in a rhythmic way, joining now the tradition of Armstrong, Young, and most of the bebop world. Clearly, his later comment that he "tried to play bebop . . . but it was too hard" (Tesser 1980: 17) was merely a jest, for he was playing mainstream bebop in the 1950s, albeit in a gentle fashion.

In 1961 Konitz recorded one of his most significant albums, *Motion.*[73] It is a trio album with bassist Francis "Sonny" Dallas and drummer Elvin Jones. The pairing of Konitz's gentle style with Jones's usually boisterous style (see page 189) seems strange, but the results are highly successful. Jones did most of the accommodating, by playing quietly while maintaining his busy intensity. In *Out of Nowhere* he plays with brushes in a most creative manner. Konitz plays almost continuously during the hour of music contained on the CD reissue; it is a merciless exposure of his creative thinking processes. Although he was almost surely improvising throughout the album, he does use some figures that he has used before. Among them are chromatic ascents and descents, a quartatonic figure—

—and Figure 5B (see p. 32). He also adopts in a limited way the common bebop habit of quoting outside themes. The actual themes of the pieces he is playing appear only sketchily until the final choruses. Among the best of his solos are *All of Me, You Don't Know What Love Is* (one of his best ballad recordings), and *Out of Nowhere.*

In the 1960s his tone became much richer, warmer; in his lower register he almost achieved a tenor saxophone quality, as Ira Gitler (1966: 255) observed. Perhaps influenced by Coltrane and other younger players, he sometimes played extended solos of 40 minutes or so in public performances. He also flirted for a time with the post-bebop music of Paul Bley, Carla Bley, and others. His remarkable set of duets with a variety of instrumentalists[74] includes some abstract moments in which the free meter, vague sense of tonality, and elastic treatment of the chorus structure take the music beyond the limits of bebop. Other duets—*Struttin' with Some Barbecue* with valve trombonist Marshall Brown and *Tickle Toe* with tenor saxophonist Richie Kamuca—are unorthodox only for their lack of a rhythm section. These two pieces even include unison statements of the classic solos by Armstrong and Young from the original recordings.

In recent years he has altered his intonation; pieces such as *I Hear a Rhapsody* from the duet album with pianist Michel Petrucciani seem to be studies in out-of-tune lyricism. However, once the ear adjusts to his particular family of pitches in the 15-minute versions of *'Round Midnight* and *Lover Man,* the beauty of his music shines through clearly.[75]

Lee Konitz's early style provided the point of departure for several subsequent players. "Konitz's kids" include Clifford "Bud" Shank, Ted Nash, Gary Foster, Arne Domnerus, baritone saxophonist Lars Gullin, and, most prominent, Paul Desmond (1924–77).

Desmond spent most of his career with pianist Dave Brubeck; he made his first recordings with Brubeck's octet in 1948, and was a charter member of the extremely popular Dave Brubeck Quartet from 1951 through 1967. The two reunited intermittently for concert tours in the 1970s, the last taking place just a few months before Desmond's death.

His recordings of the early 1950s reveal the essentials of his style, which remained largely unchanged throughout his career. He had a light, sweet tone quality similar to Konitz's, especially on his timidly played high notes, and many of his improvised phrases resembled closely those of his early role model. But his syncopated rhythms and frequent use of upward lip slurs and pitch bends kept him more in the mainstream of bebop. He filled his improvisations with clear diatonic sequences.[76] He consistently created tasteful, melodic solos and was adept at mining gold from even the dullest songs.[77]

His best early solos are from extended concert performances. His solo in *All the Things You Are* begins in a subdued, offhand manner and gradually builds in intensity.[78] In the fourth and sixth choruses he plays duets with himself by playing melodic fragments alternately in two different registers. In the fifth chorus he builds extended phrases with the three-beat rhythmic pattern that Konitz also liked (see above).

Throughout the solo Desmond benefits greatly from Brubeck's sensitive and interactive comping.

In another series of college concerts Desmond produced gems such as *Don't Worry 'bout Me, Le Souk,* and a fine blues, *Balcony Rock.*[79] The latter, a nine-chorus solo, is a representative sampling of his strengths and his main weakness. The first one and one-third choruses, unified by the rhythm ♩ ♩ ♩ ♩, is simple, effective, logical. The second chorus ends with a nice phrase that elicits murmurs of appreciation from the audience:

The expressivity of the phrase is due in part to the upward slurs in measures 10 and 11, but also to the subtle use of behind-the-beat phrasing. By this time most of his colleagues were using this rhythmic device that Lester Young pioneered; it is rare, however, to hear a two-measure segment played exactly one-sixth of a beat "late," as in the last of the example above. But after these nice moments have ended, he moves into sixteenth-note rhythms that contain some musical watertreading—aimless noodling around the tonic note. Desmond was not at his best in rapid passages, and he knew it; he referred to himself sardonically as "the world's slowest alto player," and said, "I tried practicing for a few weeks and ended up playing too fast" (Lees 1983: 4). Returning to rhythms with which he is more comfortable, he slips in a portion of *My Darling, My Darling.* This quote he follows with an interesting duet with himself in the fifth chorus (as he had done in *All the Things You Are,* a few months earlier):

He constructs the following beautiful chorus upon the descent of a tenth, shown here:

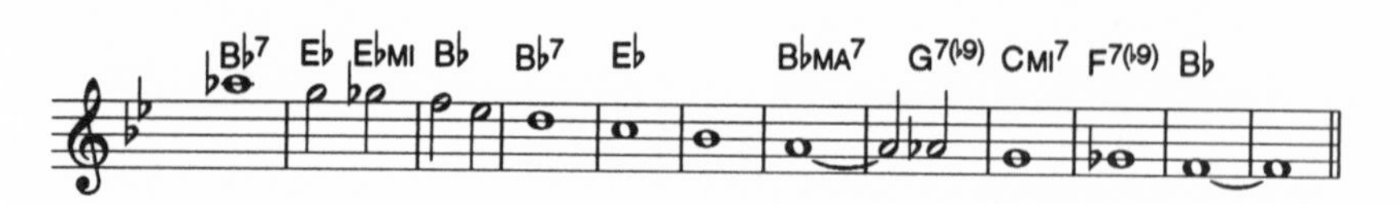

The seventh chorus he fills with pitch-bending expressivity, and then moves into another one and one-half choruses of sixteenth-note passages. These quick notes are more successful than the earlier ones, largely because of the double-time cross rhythms (1½-beat patterns repeated), one of which is similar to the Konitz phrase discussed earlier. Pausing for a breath in the middle of the ninth chorus, he hears Brubeck play a figure, which he picks up, extends beautifully, and ends the solo gently. It is a wonderful solo, the creative lapse in chorus 3 notwithstanding.

In 1959 the Quartet recorded an album of pieces written using meters that were rare or unknown in jazz: *Time Out.*[80] Brubeck wrote six of the pieces; Desmond wrote one. That one, *Take Five,* became the most popular piece in their repertoire; it even entered the mainstream of American popular music. Desmond's polite solo over a two-chord ostinato in $\frac{5}{4}$ was by no means one of his best, but the general public did not mind. And although there had been a few jazz pieces in asymmetric meters before *Time Out* appeared, *Take Five* and the other pieces in the album opened the door allowing $\frac{5}{4}$, $\frac{7}{4}$, $\frac{2+2+2+3}{4}$, and other uncommon meters into the mainstream of bebop.

Although Desmond's career was tied to Brubeck's, he recorded several albums on his own during the years with Brubeck. After the Quartet disbanded in 1967, Desmond became largely inactive musically. The albums he made in the late 1960s and the 1970s show him to have retained his sound and melodic imagination until shortly before his death. Only his final recording with Chet Baker reveals the physical debilitation of his fatal lung cancer.[81]

CHAPTER FIVE

Tenor Saxophonists

During the 1940s and 1950s, bebop alto saxophonists could have been listed as "Charlie Parker and others," but tenor saxophonists had no comparable *single* influence. Instead most tenor players drew in varying degrees from three players: Parker and two tenor greats of the swing style, Coleman Hawkins and Lester Young.

Coleman Hawkins (1904–69) had been the primary role model for tenor saxophonists of the 1930s, and continued to influence young players in the 1940s. During his tenure with the Fletcher Henderson band (1923–34) he recorded extensively, developed his mature style, and built an international reputation. From 1934 to 1939 he lived in Europe, enjoying the adulation of the jazz audience there and further solidifying his reputation. When he returned to the U.S. in the summer of 1939, most of the jazz world regarded him as the finest tenor saxophonist in jazz, and nearly every tenor player (and many alto and baritone saxophonists as well) in the profession had copied elements of his style. In the fall he made his most famous recording, *Body and Soul.*[1]

Hawkins used two contrasting approaches to improvising. In most cases he used 1) strings of swing-eighth notes, with light on-the-beat breath accents and heavy off-beat tongued accents at phrase endings; 2) a loud and full tone; 3) a raspy tone quality, produced by growling into the instrument, at climactic moments; 4) an automatic and pronounced vibrato; 5) harmonically explicit improvised melodies; and 6) a dramatic, excitement-building approach to presenting a solo statement. But sometimes in slow pieces he used a rhapsodic, florid style, filling phrases with short notes whose precise time values defy accurate notation, and playing with a quieter but still rich tone.[2]

Lester Young (1909–59) made his first recordings in 1936, shortly after joining Count Basie's band. By then he had fully formed his elegant style, as he showed with his classic solo on *Oh, Lady Be Good*[3] and shorter solos from the same session. These and subsequent recordings with Basie's band and with small recording groups led by Teddy Wilson and Billie Holiday made it clear that Young was the antithesis of Hawkins: his economical melodic lines with their unpredictable rhythms and patterns of accents, his soft, gentle, and sweet tone quality, his subtle and sparing use of vibrato, his harmonically vague improvised melodies, and his relatively undramatic approach to his solos all made him seem the outsider among saxophonists at first. But his influence grew during the 1940s, and eventually exceeded Hawkins's, particularly in the realms of tone quality and rhythm. Young's most famous recording, *Lester Leaps In,* is still a basic source of ideas for tenor saxophonists playing the *I Got Rhythm* chord changes.[4]

From the mid-1940s on, Hawkins and Young often played with younger bebop musicians, but their musical tastes were so solidly grounded in the swing idiom that neither man altered his style in any fundamental way. Other swing tenor players did change, however, and became transition figures. Budd Johnson (1910–84), for example, established his career well before the advent of bebop. But he wrote modern arrangements for the Earl Hines and Billy Eckstine bands in the middle 1940s, and gradually modified his playing style to incorporate newer melodic ideas. Similarly, Don Byas (1912–72) was a well-established member of the Coleman Hawkins school before the first bebop recordings appeared. And while he never changed his tone quality or use of vibrato in any fundamental way, he did change his melodic vocabulary in the mid-1940s. In *I Got Rhythm, I Found a New Baby,* and other solos his long, eighth-note-dominated phrases are closer to bebop than to swing.[5] They include several figures associated with Parker—Figures 3, 4E, 5A, 5B, 9, and 11B (pp. 31–33)—and have accent patterns typical of bebop. He took part in several of Dizzy Gillespie's record dates in the 1940s and 1950s; during the 1960s he used Bud Powell, Kenny Clarke, and other beboppers on some of his record dates, and occasionally recorded bebop tunes. By that time his style was essentially that of a bebop player.

Eli "Lucky" Thompson (born in 1924) was more clearly aligned with bebop. Although his best-known 1940s recording, *Just One More Chance,*[6] is a swing-style Hawkins-inspired ballad all the way, he was flexible enough to participate in some key bebop recordings: Parker's *Yardbird Suite* and *Ornithology* of 1946, Gillespie's *Confirmation* of 1946, and Thelonious Monk's *Skippy* of 1947. He gradually adopted much of the newer vocabulary, and by the 1950s was more of a bebop than a swing player, as he proved handily in Miles Davis's *Walkin'* of 1954 and Milt Jackson's *Second Nature* album of 1956. By the 1960s his tone quality, phrasing, and use of vibrato, on both tenor and soprano saxes,[7] were far from the swing-style norm. His contributions ended in the 1970s, when he retired from performing.

The first tenor sax players to gain reputations as bebop players were Dexter Gordon (1923–90) and Gene Ammons (1925–74), partly because they played in Billy Eckstine's bebop big band of 1944–45. Gordon was a central figure in the Los Angeles jazz world. A participant in recording sessions with Eckstine, Gillespie, Parker, Powell, and Roach during 1944–46, he had an impact on several young bebop tenor saxophonists.[8]

Gordon borrowed heavily from Lester Young's style in the forties, making much use of the older man's phrasing and articulating habits, his inverted mordent on F (Figure 2A from page 31), his favorite cadential formulas, and his use of alternate fingerings of B♭ to get a repeated-note effect. But he used high notes that Young never used and had a less mellow tone quality. Further, Young's melodies did not have the strings of separately articulated eighth notes, or the chromatic phrases that Gordon played at times. He borrowed some ideas from the Hawkins school as well, such as the romantic, breathy, sixteenth-note passages in *So Easy*.[9]

At this time he was largely unaffected by Parker, though he did use some basic figures that Parker, and he, probably learned from swing-style players (i.e., Figure 5A from page 32), and some Parker-like sixteenth-note flurries in one or two pieces. His melodies lack the harmonic clarity that Parker's regularly have in abundance; for example, in measures 8–11 of the blues in B♭, an area in which Parker's harmonic intentions are usually crystal clear, Gordon's harmonic choices seem almost intentionally vague.[10] They also lack the subtly disguised scalar descents of Parker's lines; Gordon's descents generally run less than an octave and lie open-faced on the surface, sparsely garnished by a few chordal leaps and neighbor tones.

One figure, based on the rhythm ♩ ♫ ♩, became a Gordon trade mark; in *Long, Tall Dexter*, take 2, it took this form:

Another Gordon lick appears in *Dexter Digs In,* take 2, from the same session:

This figure, which Don Byas also was using at the time, nearly always ends on an altissimo note, made especially dramatic by a slight delay in arriving.

In *Dexter's Deck,* from the 1945 *Blow, Mr. Dexter* session, another Gordon trait surfaces when he plays slightly behind the beat in this bright-tempo piece. Young, the pioneer in this technique, was using it only in slow pieces at the time. Gordon began walking that rhythmic tightrope early, and did so in nearly all tempos.

Some bootleg recordings from the 1940s and early 1950s reveal the ham in the man. Following the lead of such grandstanding players as "Illinois" Jacquet, Gordon would play simple, repetitive patterns designed to elicit approval from the audience; he also would quote themes in a blatant, unsubtle way so that the joke was obvious to all.

The 1950s were turbulent years for Gordon, as he dealt with the drug problem that plagued many players during that time. Between 1952 and 1960 he had only three record dates. But in those dates he announced clearly that he had severed his ties to the swing era. His solos on *Daddy Plays the Horn, Confirmation, Number Four,* and *You Can Depend on Me* are inspired bebop improvisations that mark the beginning of his mature style.[11] The melodies are much more explicit harmonically than before; they include such basic bebop ingredients as Figures 3, 5B, and 7 (pp. 31–2), and longer structural scalar descents. Obviously the Parker vocabulary had made a deep impression on him in the early 1950s. He quotes Horace Silver's 1953 bebop theme *Opus de Funk* near the end of the long *Daddy Plays the Horn* solo. Complementing his by-then traditional bebop vocabulary are 1) long phrases of even, separately articulated eighth notes, 2) a slight lagging behind the beat in almost all pieces, 3) an abundance of upward lip slurs, 4) a sprinkling of tune quotations, and 5) an important new way of playing Figure 1B (page 31), with the first note of the arpeggio lengthened:

In the ballad *Darn That Dream,* from the same album, he uses two of his favorite theme embellishments. First, he makes two notes out of one (the Ds, $E^{\flat}$s, and Fs), creating a melody that would be sung "Da-arn that dre-eam I dream each ni-ight":

Second, he embellishes a long note with an upper neighbor tone, using the characteristic articulation shown in this example (measure 13 of the theme)—

From 1960, when his personal problems smoothed out, through the mid-1980s, when his health began to decline, he played and recorded excellent bebop. In 1960 he announced his return to recording with an album that included *Lovely Lisa,* a medium-tempo piece that confirmed his firm command of the idiom. The following year he began an important recording series for Blue Note Records. The first high point in this series is *You've Changed,* a treatise on the art of ballad playing.[12] The descending chromaticism of the melody has the potential for cloying sentimentality, but Gordon caresses the notes with a gorgeous, full-bodied tone and recasts the rhythms in ever-fresh and subtle ways. A notated example here would be useless; the solo must be heard to be appreciated.

The next album, *Dexter Calling,* does not contain any one outstanding solo,[13] but the rhythm section of pianist Kenny Drew, bassist Paul Chambers, and drummer Philly Joe Jones is superb, especially in *Modal Mood,* and the beautiful, clean sound of the recording is up to recording engineer Rudy Van Gelder's usual high standards.[14] In a later session, however, he created a fine energetic solo on his 56-measure tune, *Cheesecake.*[15]

In 1962 Gordon went to England and the Continent. Except for brief return visits to the U.S., he lived and performed in Europe (Copenhagen mostly) until the late 1970s. While in Paris he recorded an album with Bud Powell, bassist Pierre Michelot, and Kenny Clarke[16].

His best solos are in *Scrapple from the Apple, A Night in Tunisia,* and in the ballads *Stairway to the Stars* and *Willow, Weep for Me.* Gordon's rhythmic brinkmanship stands out in the latter piece; how can he lag behind so much and still play with the rhythm section?

Gordon begins his *A Night in Tunisia* solo with an effective, logically constructed four-measure break. After the break is a phrase he perhaps derived from the opening measures of John Coltrane's famous *So What* solo (discussed below). Later in the chorus are some variants of Figure 1B (page 31), which he and Coltrane both began recording in 1955. Upward lip slurs to long notes played without vibrato—a Coltrane trait—appear here and there. Some rapid sweeps near the end of the first and second choruses must have been inspired by Coltrane's "sheets of sound" effect (discussed below). Throughout the solo the intense tone quality reminds the listener of Coltrane's sound. And in the closing cadenza Gordon uses the diminished scale, and even plays Coltrane's favorite diminished-scale run (illustrated below). Thus, one of Coltrane's early role models is now imitating the imitator.

These borrowings from Coltrane's vocabulary occurred often in Gordon's later solos. In one case the borrowing extended beyond phrases and articulations; beginning in the 1960s, Gordon used portions of Coltrane's unique reharmonization of *Body and Soul* (see below). But Gordon never confined his borrowings to any one role model; he took what he liked from various places, adding new elements to whatever was already in place, and ingeniously preserving his musical identity with great clarity. The borrowings were from stylistically disparate sources: Hawkins, Young, Parker, Coltrane, and others. One of those others was Ben Webster, who, in the middle of his famous *Cottontail* solo[17] applied an exaggerated "lip trill" to the note A♭ (fingered as B♭ on the tenor saxophone), producing a wobble, or shake, between A♭ and the F below it.[18] From at least the early 1960s on, Gordon used this "Webster wobble," generally in fast pieces where A♭ is the fifth or lowered seventh in the key. He was not the only borrower of this effect, for it is the common property of many swing and bebop tenor saxophonists. But he perhaps used it more than anyone else.

His next Blue Note recording, also done in Paris, contains a performance of *Darn That Dream* that surpasses by far the version of nine years earlier.[19] It is the best-recorded piece from the session, as well; the others have a very poor mix of instruments (Rudy Van Gelder's mastery is sorely missed here).

Many of Gordon's best performances date from concerts in clubs and concert halls. The intensity of his solo in the 1967 *Sonnymoon for Two,* recorded at the Jazzhus Montmartre in Copenhagen with Kenny Drew, bassist Niels-Henning Ørsted Pedersen and drummer Al "Tootie" Heath, would be difficult to duplicate in the studio environment. Similar comments apply to his 1975 *Tenor Madness* and others.[20] Some of these solos seem overlong, with too many tune quotations and predictable phrases, such as the following:[21]

But after experiencing the joy of hearing him teeter on, but never fall over, the brink of rhythmic disaster, as he chases the rhythm section a fraction of a beat "late," we can overlook any melodic shortcomings his solos may contain.

You've Changed, from the 1975 concert recording, is another expressive ballad performance. Before playing, he recites in his unique, slow-motion voice the lyrics for the first eight measures. The recitation reveals much about the man and his music; he simply marched (or ambled) to the beat of a different drummer. And a slow drummer it was when the tune was a ballad. In the 1970s and 1980s he favored tempos of ♩ = 55 and slower; *Easy Living*—with a tempo of ♩ = 38!—may belong in the *Guinness Book of World Records.*[22]

In the late 1970s he moved back to the U.S. The good recordings continued: an interesting album in an uncharacteristic (for him) eleven-man ensemble; a beautifully recorded album featuring a fine version of *Body and Soul;* a fine, extended jazz-club performance of *Come Rain or Come Shine;* a videotaped concert performance at Iowa State University, which features some fascinating side-view closeups of Gordon playing; and perhaps his finest ballad recording, *Skylark.*[23]

In 1985 he starred in the film *Round* (sic) *Midnight,* the story of a fictionalized American jazz musician, down and almost out, living in Paris around 1960. Elements of the story derive from the lives of Bud Powell and Lester Young. Gordon's background gave him many insights into the character he portrayed, and he is fascinating to watch. No other actor could learn to walk, talk, and move as he does without lapsing into an awkward caricature. The film's director, Bertrand Tavernier, wisely filmed and recorded simultaneously the extensive musical scenes, creating an immediacy and realism that sidelining to prerecorded music cannot achieve. Unfortunately, Gordon's health at the time was unstable, and his music is below par. The playing he did on a subsequent Tony Bennett album was better,[24] but these late recordings in no way match the masterful recordings he made in the 1960s and 1970s.

Near the beginning of his career Gordon may have influenced Gene Ammons, his tenor-playing partner in Billy Eckstine's band of 1944. Or perhaps both men copied similar musical elements from their early role models of the swing era. In either case, in their feature number in the band, *Blowing the Blues Away,* their styles match closely.[25] Ammons, however, grandstands a bit more with some screaming notes and high-note fall-offs. And although the two men developed different tone qualities over the years, many stylistic similarities continue all the way to Ammons's final album, recorded 30 years later.[26] Most of the albums under his name were informal small-group sessions; several featured tenor "battles" with his frequent club-group partner Sonny Stitt.[27] They show Ammons to be a good, hard-swinging journeyman.

In the 1940s, Dexter Gordon often found himself in the company of another tenor saxophonist, Wardell Gray (1921–55). Born in Oklahoma, raised in Detroit, Gray moved to Los Angeles in 1945, where he and Gordon often had "tenor battles" in concerts. They made an interesting pair: Gordon, tall and powerful looking; Gray, so small and thin that "you wouldn't think he could get a sound out of that horn."[28] Musically, however, they were well matched, and there was no winner except the audience, which heard each man spur on the other. Some of these duet performances were recorded, perhaps surreptitiously; among them are *The Chase, The Hunt,* and *The Steeplechase.*[29] Both players owe much to Lester Young.[30] But whereas Gordon plays behind the beat much of the time, Gray plays right on top of the beat. Gordon's tone quality is open and tends toward harshness, but Gray's is mellower, even covered. Gray's vibrato is faster than Gordon's. And Gray largely eschews tune quotations.

Gray's style is hard to pin down; he seems to have drawn on Young, Stan Getz (see below), and Parker in varying proportions from piece to piece. But his fast, light vibrato came from none of these three, and the warm tone and vibrato he used in ballads was also his own.

His blues solos, such as *Twisted,* take 5, contain some of his most extensive borrowing from Parker.[31] In this piece he begins with the phrase that Parker used to begin his *Billie's Bounce* solo in 1945. The second chorus opens with a quotation from the popular song *Swinging on a Star,* the rhythm of which——he repeats four times. Parker also used this same pattern occasionally as a rhythmic ostinato, as early as 1946.[32] The rest of the solo is a nearly unbroken chain of figures used by Parker.[33] But in *Southside,* the next piece from the same session, Parker's influence fades and Getz's rises.

Gray's professional career lasted only one-fourth as long as Gordon's, and his recorded output is correspondingly smaller. In the spring of 1955, someone found Gray's body in the desert northeast of Los Angeles. His death—which followed Charlie Parker's by only two and one-half months—is shrouded in mystery, but the drugs that played a large role in his life probably played a part in his death as well.

Besides Gordon and Gray, there were other tenor players establishing themselves in bebop in Los Angeles; among them was Teddy Edwards (born in 1924). His early recordings, dating from the period 1945–47,[34] show him to be a promising player trying to find his way into bebop, using a tone quality and vibrato similar to those used by Hawkins. By the late 1950s, he had found a very distinctive style; the roughness of his earlier sound had developed a mellow veneer, and he mixed a few of the standard Parker licks (Figs. 1A, 2A, 3, 5B, 8, 9) with many blues-scale phrases and pitch inflections. All his best solos sounded like blues solos, even when the chord structure was a popular song such as *Our Love Is Here to Stay.*[35] In the 1960s he explored John Coltrane's vocabulary of range-spanning runs and vibratoless notes, especially in his solos with the Gerald Wilson big band. But that exploration proved to be temporary; his style today is similar to his style of the 1950s.

Stan Getz (1927–91) started as a Lester Young imitator and became a world-class jazz stylist. A teenaged veteran of Jack Teagarden's, Stan Kenton's, and Benny Goodman's big bands, Getz made his first small-group recordings when he was 18. In them his tone quality, rhythmic approach, vibrato, and even a borrowed phrase all derive from Young,[36] but he had a more forceful manner and lacked Young's subtle shadings of articulation and tone. The youthful Getz was not yet sure of his musical direction, for in his first session as leader a few months later his tone was much harsher. Perhaps he was spurred on by the aggressive accompaniment of pianist Hank Jones, bassist Curly Russell, and Max Roach, or perhaps he had been listening to Dexter Gordon, and was trying a mouthpiece that produced a bigger, richer tone quality than the type he eventually came to use. But the tenor saxophonist in *Opus de Bop* and *Running Water* is not the one we normally think of as Stan Getz.[37]

While Getz had received a measure of recognition in his first big-band jobs, he first won acclaim in the Woody Herman band of 1947–48. He was one of the players for whom Jimmy Giuffre wrote *Four Brothers* (see pp. 21–22). Getz's solo in that classic piece, and his delicate solo in

Early Autumn,[38] earned him a place among the principal young tenor sax players of the day, and enabled him to build a successful career as a small-group leader when he left Herman at the beginning of 1949.

Early in his new career phase he recorded *Crazy Chords*, a blues that begins and ends in E♭, with choruses rising chromatically through all the remaining 11 keys. Getz improvises in seven of the keys (pianist Al Haig has the rest), and seems as comfortable in E, F#, and D as he is in the more common keys.[39]

There is hardly a trace of the Parker style in these early recordings, but his continuing allegiance to the Young esthetic shows up clearly. He exploits the upper range of the instrument much more than Young did, however, especially in ballads, where his delicately controlled high notes sound as though they are coming from a lightly played alto sax, not a tenor. And always there are more notes per phrase than Young normally used, and there is an extra forcefulness that Young did not exhibit. Except in ballads, where his behind-the-beat playing closely approaches Young's method, he plays right on top of the beat, pushing ahead with gentle but firm resolve. *What's New* is among his finest ballad recordings in the early 1950s.[40] In this lovely performance, done with Charlie Parker's rhythm section at the time (Al Haig, Tommy Potter, Roy Haynes), Getz begins with a theme paraphrase in which he masquerades almost perfectly as Young. Then, after four measures, he jumps to the upper register and reveals his true identity. From there on the similarity continues, but there is no mistaking him for Young; the faster and wider vibrato and the abundance of high notes and ornate phrases serve to differentiate the two players.

Two excellent fast performances from this period are *Mosquito Knees* and *Lover, Come Back to Me*.[41] The first comes from a night-club performance, again with Parker's rhythm section (Haig, Jimmy Raney, Teddy Kotick, Tiny Kahn). Some traces of Young's tone and use of vibrato remain, but Getz's wonderfully energetic, fluid, and mellow improvisations are his own. Except for two or three fleeting moments Getz seems almost totally immune to Parker's influence. But in *Lover, Come Back to Me* there are more traces of Parker's vocabulary: modified versions of Figure 3 (in the third and fourth measures of several *a* sections), plus 1A, 2A, 4A–F, 7, 8, 13 (pp. 31–33), and other less common Parker patterns. These figures usually occur in isolation from one another, however; he does not connect them into longer Parker-like phrases. Further, several times he uses consecutive repetition of two-, three-, and four-note patterns, and some figures involving repeated eighth notes—all foreign to Parker's idiom. And his smooth tone quality differs markedly from Parker's.

In the late 1950s Getz's tone quality in the high register developed a new huskiness, and by the 1960s that huskiness often extended into his middle register as well. Many of his later solos consequently sound more emotional and seem to swing harder than the earlier solos. Also about this time the pitch span of his vibrato widened, almost becoming a wobble. This widening is most noticeable in the ballads where long notes are plentiful.

In the 1950s Getz became a superstar of jazz; he recorded and concertized extensively throughout the U.S. and Europe, sometimes with his own quartets and quintets, sometimes with other giants such as Oscar Peterson, Count Basie, Dizzy Gillespie, and Ella Fitzgerald. Norman Granz, who owned Norgran, Clef, and Verve Records, and was the Sol Hurok of the jazz world, orchestrated this high-profile activity. Getz earned his successes honestly, for his improvisations were consistently fine. His elegant and unerring command of the bebop language inspired one writer to describe his playing as a magical "lacing together with velvet threads" the chords of a song (Diaz 1991). Any attempt to list the best of these magically laced solos is an exercise in arbitrary selection. But here are a few:[42]

Crazy Rhythm—30 July 1953, Norgran 1000

Cherokee—1 August 1955, Norgran 1037—a quintet performance with vibraphonist Lionel Hampton

Bronx Blues, I Want to Be Happy, and *Three Little Words*—10 October 1957, Verve 8348 and 8251—with the Oscar Peterson Trio

Blues for Janet—11 October 1957, Verve 8252—with the Peterson Trio

Too Close for Comfort, This Can't Be Love, and *Scrapple from the Apple*—12 October 1957, Verve 8249 and 8348—with baritone saxophonist Gerry Mulligan (on the first piece the two saxophonists play each other's instruments)

Billie's Bounce, It Never Entered My Mind, and *Blues in the Closet*—19 October 1957, Verve 8265—a concert performance with trombonist J. J. Johnson and the Peterson Trio

Evening in Paris—February 1961, Verve 815 239-1—with the rhythm section of the Adderley Quintet (Victor Feldman, Sam Jones, Louis Hayes)

Night Rider from *Focus*—1961, Verve 8412—a work composed by Eddie Sauter for jazz soloist and strings

Samba de Uma Nota So (One-Note Samba) and *E Luxo So*—13 February 1962, Verve 8432—from the popular *Jazz Samba* album with guitarist Charlie Byrd

My Heart Stood Still—April 1964, Verve 8833—with Bill Evans, Richard Davis, Elvin Jones
Summertime—19 August 1964, Verve 8600—recorded in concert when his group included vibraphonist Gary Burton
Stan's Blues—9 October 1964, Verve 8623—recorded in concert with Gary Burton
Windows—30 March 1967, Verve 8693—recorded when his group included pianist Chick Corea
Happy 50th Stan. . .—January 1977, Steeplechase 1073—performances at the Montmartre in Copenhagen, with the superb rhythm team of Joanne Brackeen, Niels-Henning Ørsted Pedersen, Billy Hart
A Time for Love and *Joy Spring*—10 May 1981, Concord 158—recorded in concert

The energetic *Blues in the Closet* contains some bows to his predecessors: to Dexter Gordon, with two quotes of the second theme of *LTD*; to Charlie Parker, with a modified version of Figure 3, combined with Figure 5B, in the eighth measure of several choruses—

—and especially to Lester Young, with one of his favorite cadential figures—

—and one of his favorite pentatonic licks—

This solo also contains references to Getz's contemporaries, for he uses the diminished scale, a common bebop melodic device in the 1950s.

The work *Focus* by Eddie Sauter is an eclectic work, with bits of Stravinsky, Bartók, Broadway schmaltz, and other ingredients mixed together in the string parts. Getz is the soloist almost continuously in this seven-movement work; working from a sketch of the string parts, he improvised everything he played, often taking motives from the string parts and developing them. The orchestra's performance suffers from inadequate rehearsal, but the result is intriguing nonetheless.[43] In the opening movement (which is actually two takes of the same movement spliced together) Roy Haynes, Getz's drummer at the time, contributes some effective brush work.

The *Jazz Samba* album was an influential recording, for it triggered a popular interest in blending jazz and Brazilian music into the *bossa nova*—a Portuguese term for new songs performed as slow sambas. At the time the song *Desafinado* was the best-known piece in the album, but the two on the list above are better. This album inspired Getz (or Norman Granz) to produce several more albums of *bossa nova* music with Brazilian musicians, among them Antonio Carlos Jobim, Luiz Bonfa, Joao and Astrud Gilberto, and Laurindo Almeida. For several years Getz and *bossa nova* seemed synonymous.

During the last three decades of Getz's career he continued along the lines established in the 1950s: worldwide tours with a variety of small groups (mostly quartets), appearances at jazz festivals with all-star groups, albums in a variety of musical settings. His style remained unchanged and his abilities undiminished. In the fall of 1987 Getz had major surgery, which sidelined him for a time. But after a period of recuperation he continued to perform regularly until shortly before succumbing to cancer of the liver.

Two of Getz's section mates in the Woody Herman band of 1948, John "Zoot" Sims (1925–85) and Al Cohn (1925–88), were also Young's disciples in the 1940s and 1950s. All three sounded much alike at first. Sims's style remained comparatively close to Getz's throughout his career, while Cohn's tone quality roughened considerably over the years, approaching that of Dexter Gordon. Sims and Cohn both recorded extensively and made concert tours and albums together.[44] They swung unfailingly and played excellent mainstream jazz for decades.

Warne Marsh (1927–87) was the main tenor saxophonist in Lennie Tristano's sphere of influence (see pp. 22–23). His tone in the 1950s was light and airy, lighter even than Getz's at the time. He sounded more like alto saxophonist Lee Konitz than like any other major tenor player. In fact, his extraordinary command of the altissimo register enabled him to play as high as alto saxophonists. In *I Can't Get Started,* he plays a G and an A (concert pitch) above the treble staff with ease and gentleness.[45] Often dismissed as an anemic, dispassionate player, he was rhythmically ingenious and could weave single motives into long, imaginative melodies. In his later years his tone acquired a huskiness, but he always maintained a deceptively tentative sound.

The players discussed so far all began to establish their reputations during the 1940s. In the 1950s a slightly younger group of players, building upon the foundations laid by their predecessors, began expanding the vocabulary.

Theodore "Sonny" Rollins (born in 1930), a native of New York City who grew up in a jazz milieu, began playing professionally with Jackie McLean, Thelonious Monk, Max Roach, Miles Davis, and others while still in his teens. He made his first records in 1949 on dates led by Babs Gonzales, J. J. Johnson, and Bud Powell. For the next five years he recorded sporadically, mostly as a sideman. During a noteworthy session with Miles Davis on 22 June 1954, three of the four tunes recorded were his compositions. All three became well-known jazz standards: *Airegin, Oleo,* and *Doxy.*[46] In these pieces and in others dating from the mid-1950s he showed clearly that he was a first-rate bebop tenor saxophonist, deserving admission to the group of major players that included Gordon and Getz.

Rollins's first records reveal a player heavily influenced by Charlie Parker's phrasing, melodic ideas, and tone quality. But by 1951 a distinctively "covered" tone quality appeared, a tone that has marked his style ever since. Pieces from those early recordings show not only Parker's influence but that of others as well: a touch of Getz's or Sims's tone in *Dig,* a bit of Young's gentle approach to ballads in *No Line,* and some of Gordon's phrasing and sense of humor in *Tenor Madness.*[47] But he had a taste for un-jazzy themes that other beboppers had not considered previously: *Shadrack, All of a Sudden My Heart Sings* (renamed *Silk 'n' Satin*), *There's No Business like Show Business, How Are Things in Glocca Morra?, I'm an Old Cowhand, Wagon Wheels, Toot, Toot, Tootsie, The Last Time I Saw Paris.*[48] And in some pieces—*Reflections, Blue Seven, Ee-ah, Wonderful! Wonderful!, Blues for Philly Joe, Sonnymoon for Two,*[49] and a few others—he used thematic material throughout solos or developed single motives over extended sections of solos. In the mid-1950s he and pianist Thelonious Monk were almost the only jazz musicians who used these procedures.

His justly famous motivic improvisations in *Blue Seven* were examined in loving detail by Gunther Schuller (1958), shortly after the record appeared. In his article, Schuller explains that much of what

Rollins plays in his fourteen full choruses and three partial choruses (shared with drummer Max Roach) derives from the first three measures of the theme and from the three pick-up notes leading into his second improvised chorus:

The swing-eighth-note motive of the theme, with its emphasis on E (the diminished fifth or augmented eleventh of the tonic chord) is one that Rollins found attractive in 1956; he used it in *Strode Rode* and *You Don't Know What Love Is* from the same recording session and in *Vierd Blues* and *When Your Lover Has Gone* from earlier sessions.[50] But he used it more extensively and creatively in *Blue Seven* than in any other recording.[51] While Schuller and other music scholars marveled over this classic performance, Rollins was surprised by their descriptions of his solo: "I was just playing what I know and I didn't know I was doing that [i.e., improvising thematically]" (Blancq 1977: 102). But consciously or not, Rollins created a classic jazz performance, one that should be heard by any serious student of bebop.

The *Blue Seven* session was especially productive for Rollins; he also recorded two more of his compositions, *Strode Rode* and *St. Thomas,* which represent two widely contrasting styles of improvising. *Strode Rode,* with a structure of 12 + 12 + 4 + 12 measures, is an intense, minor-mode performance, while his now-famous *St. Thomas* is a lighthearted romp through some basic chord changes. It is perhaps the best-known example of his calypso music, which is another distinguishing facet of his style. In these pieces (which also include a later version of *St. Thomas,* plus *Brownskin Girl, Hold 'em, Joe, The Everywhere Calypso, Don't Stop the Carnival, Little Lu,* and *Duke of Iron*[52]) there are usually sections built upon the basic I and V^7 chords; during these sections his improvisation is correspondingly rudimental, based almost entirely on a single motive in the major scale.

Rollins is justly famous for his unaccompanied playing, such as his 1958 recording of *Body and Soul.* Throughout these two ornate paraphrase choruses he weaves in and out of various tempos with a spontaneity that would have been impossible had the rhythm section been playing. Although he frequently performs an unaccompanied solo in concerts, his unaccompanied moments on record usually are during introductions and codas. One example is in his beautiful but agitated version of *Skylark*. In the coda, which is really an extended cadenza, he plays an entire chorus of the melody, with many interpolations of other themes and many tempo changes.[53]

Rollins's tone quality, perhaps his most readily identifiable characteristic, is difficult to describe in words and impossible to show in notation. Writers have called his tone "cello-like" (Williams 1970: 143), "hard, sometimes to the point of deliberate harshness" (Feather 1960: 401), "rough," "brittle," "coarse and guttural" (Gridley 1991: 208–9). Whitney Balliett said Rollins has a "bleak, ugly tone—reminiscent, at times, of the sad sounds wrestled by beginners out of the saxophone. . . . He seems . . . to blat out his notes as if they were epithets." Elsewhere Balliett says Rollins has a "goatlike tone" (1959: 45, 70). Bill Green, in talking to me about Rollins's self-taught, devil-may-care playing techniques, called him a "diamond in the rough."

His tone is hard to describe partly because he uses different sounds for different types of pieces—a feature, to be sure, found in the playing of Coleman Hawkins, Ben Webster, and others. "Cello-like" creates an aural image of rich, deep, mellow sound—a facet most clearly heard in ballad recordings such as *How Are Things in Glocca Morra?, The Most Beautiful Girl in the World, Body and Soul,* and *Stay as Sweet as You Are.*[54] "Harsh," "rough," and "hard" suggest a quite different sound, one more typical of his faster performances.[55] The unusually percussive attack that he often uses on the first notes of short phrases may have inspired these adjectives. However, his sound is no harsher or rougher than that of Parker, Gordon, and others of the time. In fact, the harshness is somewhat muted, as though a cloth were covering the bell.[56]

On 3 November 1957, at the Village Vanguard in New York City, Rollins made some of the most important bebop recordings of the 1950s. The three LPs from that evening document two different saxophone-bass-drums trios. Rollins had previewed his interest in smaller ensembles as early as 1955; in *Paradox,* he exchanged four-measure phrases with Max Roach, with no other instruments taking part. His first trio album, *Way Out West,* with bassist Ray Brown and drummer Shelly Manne, dates from early 1957. But the Village Vanguard recordings, which capture Rollins in the heat of inspiration stimulated by an audience, are special.[57]

For most of the pieces the bassist is Wilbur Ware and the drummer is Elvin Jones; in two pieces Donald Bailey and Pete LaRoca replace them. The bare-bones two-part counterpoint of tenor sax and bass places heavy demands on Rollins, who, spurred on by the exciting drumming of Jones or LaRoca, consistently rises to the occasion in this series of spectacularly creative performances. His fertile imagination surfaces dramatically in *Softly, As in a Morning Sunrise, Get Happy,* and in the two almost entirely different solos on *A Night in Tunisia.* His exceptional command of time in the closing theme of *A Night in Tunisia* (the version with Bailey and LaRoca) is a joy to hear. He plays phrases in an almost unmetered way, but brings them to a close at just the right moment. And the icing on this musical cake is the beautifully clear sound mix produced by Rudy Van Gelder.

Early in 1958 Rollins recorded another major work, *The Freedom Suite.*[58] Again the group is a trio, this time with Oscar Pettiford and Max Roach, and there is much creative interplay among the three musicians. This 19-minute work contains several subsections in a variety of tempos and moods. A short gospel-like waltz serves to link three longer sections.[59] The work is a statement about the injustices of racial prejudice that Rollins had experienced, and a reminder of the enormous contributions that black Americans have made to the culture of America as a whole. Rollins wrote the following statement for the liner notes:

> America is deeply rooted in Negro culture: its colloquialisms, its humor, its music. How ironic that the Negro, who more than any other people can claim America's culture as his own, is being persecuted and repressed, that the Negro, who has exemplified the humanities in his very existence, is being rewarded with inhumanity.

Another important trio album dates from a European radio broadcast in 1959, with bassist Henry Grimes and drummer Pete LaRoca.[60] Among the gems here is a fine performance of *St. Thomas* (marred by one rhythmically confusing moment in the fours between Rollins and LaRoca), a lovely, warm performance of *Stay as Sweet as You Are,* an impish arrangement of *I've Told Every Little Star,* and a rhythmically inventive *How High the Moon.* A more varied jazz repertory would be hard to imagine.

In the fall of 1959 he voluntarily stopped playing in public, to deal with some personal problems and to re-examine his musical habits. During his more than two-year leave of absence he worked hard on his playing style, often practicing at night on the Williamsburg Bridge over the East River in New York City. Some recordings made after his return to professional life, such as an excellent version of *The Night Has a*

Thousand Eyes, show little obvious change in his playing; others, such as his album with Don Cherry, Bob Cranshaw, and Billy Higgins, show him going "outside" to dabble in action jazz, something he has done on occasion ever since.[61]

In the 1970s he explored another new development in jazz—fusion. Several of his recordings during that time have strong rock influences: *Playin' in the Yard, Lucille, Camel,* and others.[62] He also played the Lyricon, a synthesizer controlled by a wind instrument, on a few recordings.

In the 1980s and early 1990s his style is a composite of many elements: the bebop foundation, the ingenious rhythmic flexibility, the flurries of notes reminiscent of action jazz, the creative simplicity of calypso style, the rock-influenced rhythms, and perhaps above all, that one-of-a-kind sonority. His sound is a thing of beauty when properly recorded, as it is in the 1986 video documentary, *Sonny Rollins—Saxophone Colossus.*

But as great a player as Rollins was in the 1950s and is now, his impact on the evolution of jazz is less than that of John Coltrane (1926–67), who was, "after Charlie Parker, the most revolutionary and widely imitated saxophonist in jazz."[63] Earlier I discussed his impact on Dexter Gordon and others, and the even greater impact he had on younger players makes one realize that "Trane lives" just as surely as "Bird lives."

Coltrane began his career in the late 1940s and early 1950s by adopting the bebop vocabulary of the time, but soon he expanded his musical horizons. By the late 1950s he had added significantly to the possibilities open to bebop soloists, though few of them had the digital dexterity and mental agility to follow his lead at first. And when other saxophonists began catching on to his florid bebop style, he had moved into musical territories outside the bebop world. The personal quest to express himself in new ways ended tragically with his early death due to liver cancer (Thomas 1975: 224).

Coltrane was 23 when he made his professional recording debut with Dizzy Gillespie's big band of 1949 (playing *alto* saxophone), and 24 when he recorded his first tenor sax solos. There is only one short studio-recorded solo from that year, on *We Love to Boogie,* but there are several airchecks of broadcasts from Birdland during early 1951. As might be expected, traces of early influences pop up here and there: a couple of bows to Lester Young in *Groovin' High* and *We Love to Boogie;* to Charlie Parker (Figure 3) in *Jumpin' with Symphony Sid;* and to Dexter Gordon's early style, including his separately articulated eighth notes played behind the beat in *Birks' Works.*[64]

Still, he had been playing for nine years, and much of his famous late 1950s style appears in these early records: a sparing use of vibrato, especially on longer notes; upward slurs to long notes at beginnings of phrases; an intense tone quality, especially in the upper register, that suggests no other player of the time; and a descending sequential pattern (in *Jumpin' with Symphony Sid*) that precedes his favorite diminished-scale pattern (see below) by several years. His best solos are on *A Night in Tunisia*[65] and *Birk's Works*, both well crafted and confidently executed. None of them had any historical impact, however, for they first appeared on a bootleg LP long after he died.

Short tenures with Earl Bostic, Johnny Hodges, and others followed the period with Gillespie, but the first crucially important job was with Miles Davis, beginning in 1955. While with Davis, he gained international exposure and joined the ranks of the major bebop players. The first records he made with Davis, in the fall of that year, show a tenor saxophonist who has fully assimilated many elements of Parker's vocabulary, integrating them into his solos on *Ah-Leu-Cha, Miles' Theme* (better known as *The Theme*, an *I Got Rhythm* contrafact), and others.[66] Sometimes the figures sound just as Parker played them; elsewhere Coltrane combines them in ways that Parker never did, or alters one or two notes to give the figure a unique character. In addition, he takes part in the typical bebop game of theme quoting, by incorporating into his solos the beginnings of *Summertime, Bill, Pop Goes the Weasel, The Kerry Dance, While My Lady Sleeps, Humoresque*, and others. His favorite is *All This and Heaven Too*—

—which became one of the most commonly used quotations in bebop.[67]

But there is much that is pure Coltrane in his recordings from 1955–60. One feature is a result of technical imperfections in his execution. Often while playing strings of separately articulated sixteenth notes (an early Coltrane trait in itself) he tongues the new note before his fingers release the previous note, or, in playing articulated pairs of notes, he tongues the first note of a new pair before releasing the second note of the previous pair. The result is a random grace-note effect, as in this example from his solo on *In Your Own Sweet Way:*[68]

Some might argue that this effect was intentional, since it occurs in solo after solo during the mid-1950s. But the effect gradually disappeared from his style, and had all but vanished by 1960. During the late 1950s he dramatically increased his technical proficiency; this grace-note effect simply fell by the wayside as his command of the instrument and of his musical idiom grew ever more awesome.

His rapid melodies, an especially striking feature of his style, become more and more impressive as the decade wears on. In the mid-1950s he plays an unusually high percentage of double-time phrases (that is, phrases dominated by sixteenth notes rather than by eighth notes) in medium-tempo solos, sometimes forming entire choruses in sixteenths in 1956; soon double-time choruses appear in ever faster tempos. In *Black Pearls* the tempo is approximately ♩ = 170, and in *Sweet Sapphire Blues* it is ♩ = 195; yet sixteenth notes are among the longer note values![69] These solos are almost unbelievably florid, and his harmonic, melodic, and technical control is superb. At these speeds, of course, he has no chance for individual articulation of notes; hence, the grace note effect is absent, replaced by a smooth, clean fingering of scales and arpeggios.

Other characteristics have to do with tone quality. In medium and fast pieces his timbre was as hard as steel, with a large, bright, intense sound, produced, according to Lewis Porter, "through the combination of a very open (large-chambered) metal mouthpiece and a soft to medium reed, and a tight embouchure" (Porter 1983: 56). And in his upper register the intensity seems even greater, almost as if he were screaming out. In the 1960s that intensity would become almost unbearably emotional, as though he were expressing the cries of a man in deep anguish. In ballads, however, he softens that hard-edged tone with a lovely vibrato.

In looking at some of Coltrane's favorite melodic formulas of the 1950s, we see several similarities to those given for Parker (pp. 31–34). He did not favor Figure 1A, but used extensively a figure similar to Figure 1B:

The difference is that Coltrane lengthened the first note of the upward sweep and compressed the remaining notes into a brief flurry, whereas Parker usually played four sixteenth notes. Interestingly, both Dexter

Gordon and Coltrane began playing this figure at nearly the same time on records. Who copied whom? Or were both men copying a third player? Logic suggests that Coltrane was copying Gordon, for he clearly had copied the older player in his early solos. But Gordon recorded nothing between 1952 and 1955; would Coltrane in New York have known what patterns Gordon was evolving 3000 miles away in Southern California? To Coltrane this issue probably would be inconsequential; he might say gently, as he did to Frank Kofsky in 1966, that "it's a big reservoir, man, that we all dip out of. . . ."[70]

Figure 2A, used by Parker and Young, was never a big part of Coltrane's vocabulary, but when he used it he generally lengthened the first note and finished the inverted mordent almost as an afterthought (Lester Young usually did the same thing, but in slower motion):

Coltrane often used Figure 3, the descending scale built on a dominant seventh with a minor ninth. The example here shows the descent preceded by another of Coltrane's favorite phrase beginnings—a lower neighbor followed by a scalar ascent:

Of the short chromatic figures that Parker favored (Figs. 4A–F) Coltrane preferred Figure 4A, which he often extended into the complete descending bebop dominant scale, and 4C:

From here on the numbering of Coltrane's figures deviates from Parker's. This next example, Coltrane Figure 5, resembles Parker's Figure 9; Coltrane uses other rhythms to present it in addition to the one shown:

Coltrane's figure 6A, an important one in his music, resembles the decorated chromatic descent of Parker's figure 11B. Coltrane's figure 6B, a corresponding chromatic ascent, also has its antecedent in a rare Parker figure:

Like Young, Parker, and others before him, Coltrane liked to inflect the dominant chord with the augmented fifth (compare Parker's Figs. 13A and B and Coltrane's Figs. 7A and B):

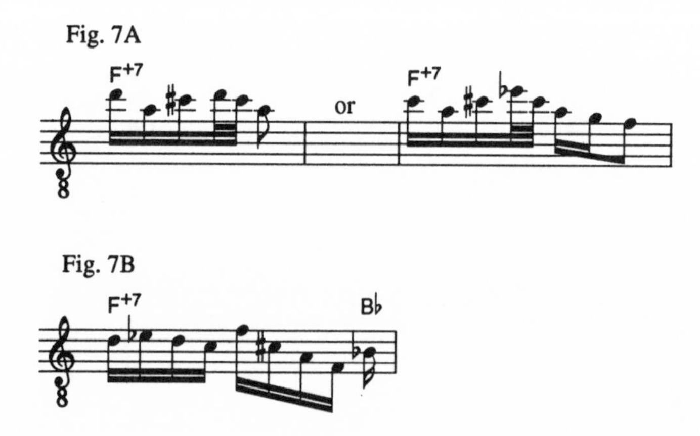

Coltrane's rapid flurry of sequences built upon the symmetrical diminished scale is an especially ear-catching figure.[71] Parker used the scale in a straightforward way in Figure 15B, but Coltrane's Figure 8 is much more distinctive:

Parker sometimes unfolded a major chord simply by playing scale degrees 1, 2, 3, and 5 of a major scale. Coltrane often did the same thing, sometimes with disturbing predictability, as Barry Kernfeld has pointed out:[72]

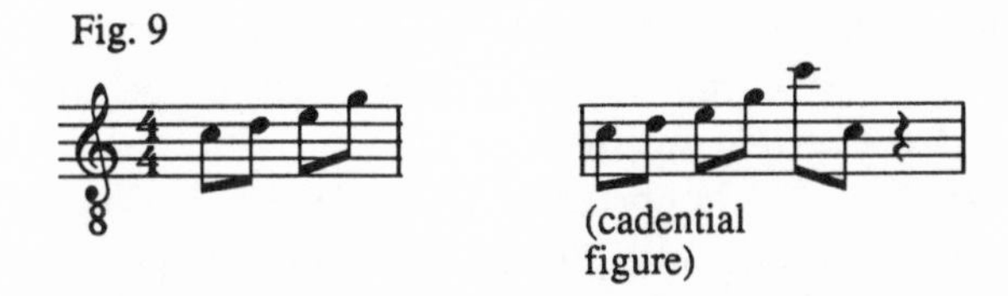

The next set of figures almost invariably occurs in passages of sixteenth or thirty-second notes; they often decorate a single chord in the harmonic structure of a song, though they appear to move through a set of chords in some cases (Coltrane may have learned Figure 10B from Stan Getz, who was using it in the early 1950s):

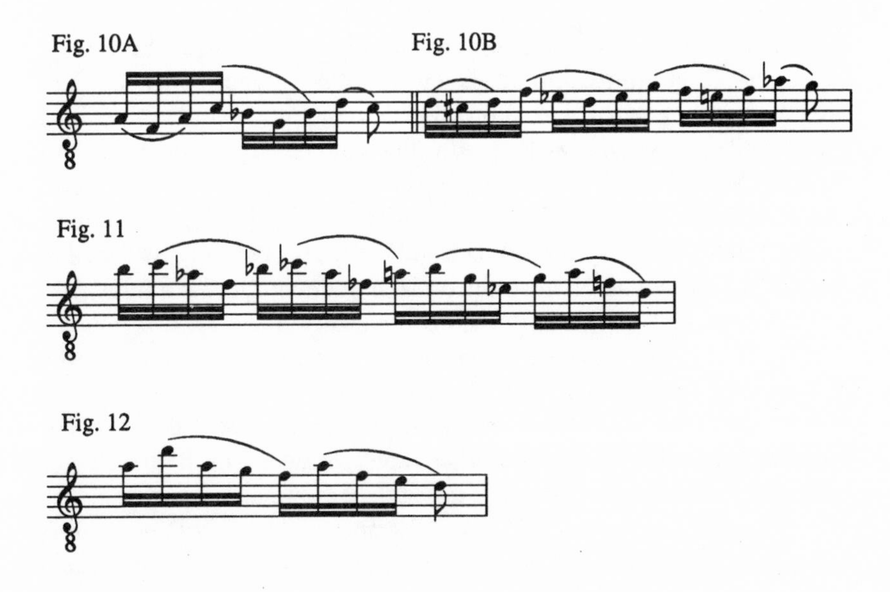

Figure 13 is an arpeggiated minor-seventh chord; it often appears, repeated insistently, over the tonic chord of a minor-mode blues or a Dorian-scale piece from the late 1950s on:

Fig. 13

This last example incorporates another important Coltrane characteristic, that of beginning a phrase with a long, vibratoless note approached either by a short upward scoop or glissando. Parker also used this simple upward slur, but seldom as the beginning of a long phrase, seldom in the high register where Coltrane favored using it, and never with the intensity of tone that Coltrane imparted to it.

While patterns are important, Coltrane's ability to integrate those patterns into a coherent style and generate memorable music is more important. And this great musician left us much memorable music to examine and enjoy.

When Coltrane joined Miles Davis's quintet in 1955, he formed a musical alliance that would have a great impact on the evolution of jazz. The best early recording sessions by the quintet were marathons on 11 May 1956 and 26 October 1956. Coltrane solos on eight of the 14 pieces recorded in May and on 11 of the 12 pieces recorded in October; his solos in *Diane, In Your Own Sweet Way, Surrey with the Fringe on Top, If I Were a Bell, Oleo, Well You Needn't, Half Nelson, I Could Write a Book,* and others form an extensive sampling of his style at the time.[73]

In 1957, he quit using drugs and alcohol, ending habits that had disrupted his personal and professional lives for several years. That spring and summer he worked with Thelonious Monk's quartet; supposedly that relationship was important to Coltrane's musical growth. Monk gave him the opportunity to play extended solos, sometimes backed by the three rhythm players, sometimes backed only by bass and drums, because Monk often stopped playing for choruses at a time. And the two men had private discussions of musical matters. But Monk's impact, if any, on Coltrane's style is hard to pinpoint. Some of the pieces Coltrane recorded with Monk's group show that he was on his way to the "sheets of sound"[74] effects of the late 1950s; they contain long upward scalar sweeps and other rapid passages. But he hardly got that trait from Monk, whose playing style was much less ornate.

During his time with Monk, Coltrane made his first session as leader, on 31 May 1957 (Prestige 7105). He recorded *While My Lady*

Sleeps, a tune he quoted often both before and after this session; *Chronic Blues* and *I Hear a Rhapsody,* which contain fine solos; and *Violets for Your Furs,* in which he reveals the great beauty of tone he could employ in ballads.

On 16 August of the same year he recorded a trio session with bassist Earl May and drummer Art Taylor, inspired perhaps by some Sonny Rollins sessions held earlier in the year (or by those piano-less solos he played with Monk's quartet). This important session produced *Like Someone in Love, I Love You,* and two takes of *Trane's Slo Blues.*[75] The move toward the "sheets of sound" effect (long passages of rapid-fire runs) is apparent in the second take of the blues. The symmetrical diminished scale seen in his Figure 8 is the basis for the entire introduction of *I Love You,* and reappears in the choruses as well. Throughout this solo, which consists almost entirely of sixteenth-note phrases, Coltrane relies heavily on his well practiced patterns.

This solo illustrates the main approach to improvisation that Coltrane followed in the late 1950s, the "mechanical formulaic" approach, to use Kernfeld's phrase (1981: 67). It occurs in most of the best-known and most frequently copied solos. The virtuosity with which he executed those formulas (or "patterns," as Coltrane called them in the Kofsky interview) was astounding and captivating, and inspired countless young saxophonists to listen intently and repeatedly to his records. But even when taped and played at half-speed, these solos present formidable challenges to students; not only do the notes fly by rapidly, but they sometimes fall into asymmetrical rhythmic groupings.[76] Coltrane talked about this aspect of his style:

> I found there were a certain number of chord progressions to play in a given time, and sometimes what I played didn't work out in eighth notes, 16th notes, or triplets. I had to put the notes in uneven groups like fives and sevens in order to get them all in. (Coltrane 1960: 27)

But there was more to his music than rapid-fire applications of well-rehearsed patterns. Kernfeld (1981: 67ff.) points out that there were three Coltrane styles at this time. Besides the formulaic one there was a second one, reserved for ballads such as the two discussed above, in which the original melodies of the songs stayed in the foreground during all or most of the solo. Sometimes he subjected these melodies to elaborate paraphrases, complete with florid passages inserted between phrases, just as in an Art Tatum ballad. In the third style, all but unknown in 1957–58, extensive developments of motives dominate.[77] This style emerged full-blown in 1959 (see below).

Early in 1958, Coltrane rejoined Miles Davis, and soon encountered a seemingly new musical context: improvising on a single scale for considerable lengths of time instead of negotiating a continuously changing stream of chords. The piece that seemed to announce this new idea was *Miles,* better known as *Milestones,*[78] whose structure is 16 measures of C Mixolydian (or perhaps F major[79]), 16 measures of A Aeolian (i.e., natural minor), 8 measures of C Mixolydian. Actually, improvisation on a scale rather than on a chord progression was not new in 1958. Bud Powell's 1951 *Un poco loco* is but one of many Latin-style pieces affording extended sections for improvisation on the Mixolydian or other scales. But when one of the world's most prominent bebop groups began to explore simple scale structures it sparked considerable interest. Largely because of the Sextet's recordings of *Milestones* and of *So What,* the bebop repertory contains many pieces calling for scalar rather than chordal improvisation.

How did this new improvisational framework affect Coltrane's approach to improvising? Hardly at all at first, for he based his *Miles (Milestones)* solo on his favorite patterns; Figures 1, 2, the first half of 3, 4A, 4C, a diatonic version of 5, 6B, and 9 all appear, some several times. The main difference is simply that he does not need to adjust them for constantly changing chords. In another piece of this type, *Oomba,* from the Wilbur Harden session of 29 June 1958, Coltrane performs in a similar manner.

The breakthrough piece of this type for Coltrane was the famous *So What* from the monumentally important Miles Davis Sextet album, *Kind of Blue* (see also pp. 57–59, 120–23). Here for the first time formulaic improvisation takes a back seat, replaced almost entirely by discrete motives spun out over segments of this piece's structure (D Dorian for 16 measures, E♭ Dorian for 8, D Dorian for 8). The main motives and their locations are as follows:

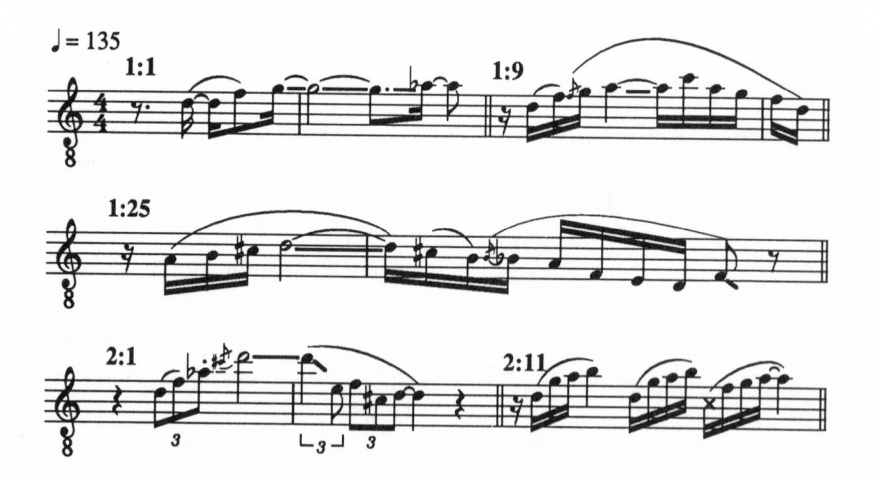

Notice that the first motive expands and becomes the second motive, that the fourth motive (chorus 2, measure 1) is in turn an expansion of the second, and that the fifth motive (chorus 2, measure 11) is essentially a horizontalization of the three-note chords that the winds play in the theme. Notice also that Coltrane leaves the strict Dorian scale and incorporates additional notes from the melodic minor scale in the third motive.[80] Finally, notice by ear the uncompromising seriousness of his performance, created by his relentless hammering away at some of these motives, his hard-edged tone quality, and the almost total absence of vibrato.

A month after this classic recording session, Davis's quintet (Adderley was absent from the group due to illness on that occasion[81]) appeared on television and played *So What* again.[82] Coltrane plays a different solo, but once again bases it on the development of a few motives. These two solos, plus similarly organized solos on *Flamenco Sketches* and *All Blues* from the *Kind of Blue* album, reveal the path that he was to explore more and more frequently during the remainder of his life.

But he was not through with formulaic improvisation in early 1959. In fact, he played some of his most rigidly formulaic improvisations on 5 May: *Giant Steps* and two takes of *Countdown*.[83] The harmonic structures of these two Coltrane compositions are similarly complex, containing overlapped III^7-V^7-I progressions:

	BMA^7 -	D^7 -	GMA^7 -	B$^{\flat 7}$ -	E$^\flat$MA7	(chord symbols)
G:	[III^7 -	V^7 -	I^7]			(chord
			E$^\flat$: [III^7 -	V^7 -	I^7]	grammar)

Coltrane became intrigued by these third-related chord successions, for he used them in his tune *Fifth House* and superimposed them on the first part of *But Not for Me* and on the bridge of *Body and Soul*.[84] These chordal relationships are significantly different from the ii-V-I patterns of most bebop tunes, and when played at tempos of ♩ = 275 (*Giant Steps*) and ♩ = 345 (*Countdown*), require some careful planning and much painstaking practice to negotiate successfully. Small wonder, then, that a close examination of Coltrane's remarkably fluent solos on these pieces reveals much recycled melodic material from chorus to chorus. Clearly he worked out his solutions to the various segments of each chord structure, and relied heavily on those worked-out patterns in performance. In the final analysis his *Giant Steps* and *Countdown* solos, made up almost entirely of long eighth-note strings, are "masterfully presented, well-planned etudes" (Jost 1974: 25).

Coltrane in the 1960s was occasionally formulaic, often lyrical, and often motivic. The lyrical, ballad-playing side came to the fore when he had some work done on his favorite mouthpiece, work that he said ruined it for playing fast (Kofsky 1970: 235). Frustrating as that turn of events must have been for him, it nonetheless resulted in two complete albums of ballads and some more ballads in a collaboration with Duke Ellington.[85] The motive-developing side of Coltrane appears in some wonderful blues recordings—among them, *Equinox*—and in some remarkable recordings of *Impressions,* a *So What* contrafact.[86] He and the brilliant, powerful drummer Elvin Jones stretched bebop to the breaking point in concert recordings of *My Favorite Things,*[87] which he played on soprano saxophone, his second instrument during the 1960s. Eventually he became a post-bebop player, as he grew increasingly interested and involved in the harmonic freedom, flexible rhythms, and intense collective improvisation used by Cecil Taylor, Ornette Coleman, Archie Shepp, Eric Dolphy, Pharoah Sanders, and others. Thus, the late-period Coltrane is beyond the scope of this book.

During the 1950s and 1960s many more first-rate tenor saxophonists came to prominence. Among this multitude is Johnny Griffin (born in 1928), one of the most aggressive, hard-playing bebop musicians. His recorded legacy from the 1950s includes some powerful blues solos; among them are *Main Spring, Terry's Tune,* and *Blues March.*[88] Early in 1958 he replaced John Coltrane in the Thelonious Monk Quartet, and almost all his recorded solos with this group are excellent. In the Five Spot Cafe recording of *Rhythm-a-ning* (Monk's *I Got Rhythm* contrafact), his blazing solo completely upstages the other players. Another night-club performance, this time with Wes Montgomery, produced some fine solos on *Blue 'n' Boogie* and *Cariba,* with the support of a superb rhythm section—Wynton Kelly, Paul Chambers, and Jimmy Cobb. In 1978 he joined Dexter Gordon for a tenor "battle" on the blues in B♭ (a traditional combat piece), *Blues Up and Down.*[89] The performance highlights the similar tone quality of the two men, but also contrasts clearly Griffin's aggressive, on-top-of-the-beat style and Gordon's lazy, behind-the-beat style. A 1985 video documentary, *The Jazz Life Featuring Johnny Griffin,* captures Griffin with a fine rhythm section in performance at the Village Vanguard.

Stanley Turrentine (born in 1934) draws heavily upon the blues and gospel traditions for both his themes and his improvised melodies.

These two brief excerpts from his well-known recording of *Sugar*—a folk-like pentatonic tune that he wrote—illustrate some ear-catching ingredients of his style:[90]

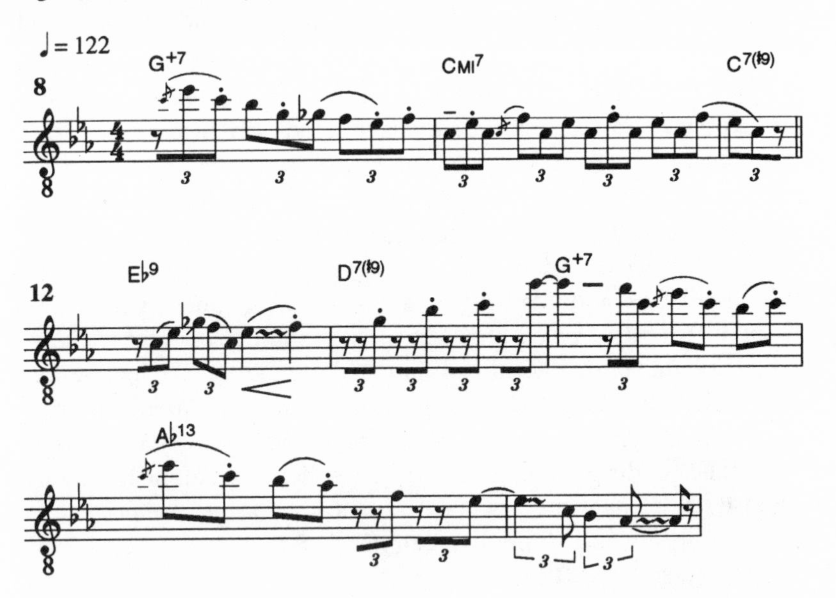

The basis of many of his improvised melodies is the minor-pentatonic scale (measure 14) and its derivative, the blues scale (measures 8 and 12). Turrentine's usage of these scales is hardly distinctive in itself, for many players use these scales to get a folk-like quality. But his manner of playing the notes is all-important. Often he tongues individual notes sharply and quickly for an accented staccato (measure 13). Just as often he will precede a sharply tongued note with a note held for its full value, resulting in clearly separated pairs of notes (measures 14–15). Also, by means of careful coordination of tonguing and fingering the first note of the pair may have a lightning-quick grace note preceding it, and that grace note is usually on the same pitch as the preceding note (in measures 14–15). Finally, there is his ready use of his nervous vibrato (measures 12 and 16) and his covered tone quality, which, while resembling Sonny Rollins's, sounds different because of the radically different context in which Turrentine places it.

Originally Turrentine leaned more heavily on the bebop vocabulary of the 1940s and early 1950s for his melodic material and used more traditional articulation habits, as in the blues *Let's Groove* and in *Mild Is the Mood* (based on the chords of Tadd Dameron's *Lady Bird*). But recordings such as *Sheri,* a gospel-tinged blues in triple meter, show that he already had much of the vocabulary that would make him famous.[91]

And everything shown in the example above also appears in his early albums on Blue Note from 1960.

Eddie Harris (born in 1934) attracted much attention with his musical activities outside the bebop mainstream. In 1961, he achieved great popular success with a bland recording of the theme from the movie *Exodus.* A few years later he pioneered in the use of electronic modifications of the saxophone, and made some early fusion recordings. He also composed the highly chromatic *Freedom Jazz Dance,* and explored unorthodox tone qualities by playing brass instruments with saxophone mouthpieces—activities more likely to be undertaken by an action jazz musician. But although he plays both fusion and action jazz with authority, he is fundamentally a first-rank bebopper. A large man with an authoritative physical presence, his sound frequently is deceptively small and tentative, his vibrato shaky; but his imaginative and harmonically explicit melodies swing unerringly. Perhaps his multiple musical interests have prevented him from gaining as great a prominence as he might have enjoyed had he followed a more traditional musical path. Though he has spent several decades in the profession, many listeners are only dimly aware of him, a fact that he laments in his humorous song *Eddie Who?*[92]

CHAPTER SIX

Trumpeters

The lineage of bebop trumpeters begins, of course, with John Birks "Dizzy" Gillespie (1917–93). His colleagues acknowledged him as a leader from the beginning, and soon the general public was aware of him. Most early articles on the new music in the popular national press focused principally upon him. His extroverted behavior and unique appearance made him much more than a trumpeter; for millions of Americans, he was the epitome of bebop. Unfortunately, the reader in the 1940s was more apt to read of Gillespie's bulging cheeks and neck, beret, horn-rimmed glasses, and goatee than to learn much about his music. Some articles, such as the one in *Time* (17 May 1948) or the photographic spread in *Life* (11 October 1948), were shallow, even silly. Others, such as the brief one in *Saturday Review* (30 August 1947) or the lengthy article by Richard O. Boyer in *New Yorker* (3 July 1948), were more informative and sympathetic, though they provided few concrete details about his art.

Gillespie's recording career began in 1937, as a soloist in the Teddy Hill big band.[1] He got the job partly because he sounded much like Roy Eldridge, who had been Hill's trumpet soloist in 1935; Gillespie's phrasing, tone quality, vibrato, and melodic ideas derived in large measure from Eldridge.[2] In several solos his use of Figure 5A (see page 32), with scale degree $\flat 2$ functioning as the $\flat 5$ of the $V^{7(\flat 5)}$, is particularly ear-catching as a borrowing from Eldridge (or Hawkins—see page 5). During 1937–42 he occasionally used the figures identified in Chapter 3 as 2A, 4A, 4B, and 4E, as well. The early recording dates suggest that Gillespie did not get these figures from Parker; but that he and Parker carried them from the swing style into the bebop vocabulary.[3] Yet Gillespie admits to being influenced by Parker: "his style . . . was perfect for our music. I was playing like Roy Eldridge at the time [early 1940s].

In about a month's time [after hearing Parker] I was playing like Charlie Parker. From then on—maybe adding a little here and there" (Lees 1989–September: 6).

Early on, Gillespie also made some independent contributions to the music. For example, in *Little John Special*[4] he played a phrase of triplets that was one of his pet phrases for several years (always in the key of B♭):

The lengthiest samples of Gillespie's early style are the famous Minton's Playhouse recordings of May 1941, mentioned in Chapter 1. Although his solos on *Kerouac* and the other pieces are not yet in the bebop idiom, they clearly show a player searching for something new. Recent evidence indicates that he found what he wanted shortly; an informal performance of *Sweet Georgia Brown,* with Charlie Parker playing tenor sax and Oscar Pettiford on bass, shows that Gillespie's basic style was nearly fully formed by 1943.[5]

In 1944 Gillespie began documenting his bebop style in a series of studio recordings, as a sideman for Coleman Hawkins, Billy Eckstine, and Sarah Vaughan. The first was *Woody 'n' You,* discussed in Chapter 2. The chords in three of the eight measures of each *a* section are half-diminished sevenths (such as GMI$^{7(\flat 5)}$), one of Gillespie's (and Monk's) favorite chords.[6] The melody in bars 7 and 8 of the *a* sections incorporates Figure 5A, using the scale steps ♭9-7-8, a figure Gillespie retained from his Eldridge-derived style. In the bridge of his solo chorus he presents almost textbook examples of two favorite improvisational habits. First comes a phrase that begins on high notes and plunges downward for two octaves. Then comes another phrase of triplets, one that was a signature phrase from at least 1943 to the end of his career. He always played it at the pitch level shown here, so that he could use alternate fingerings of D and E♭ to make the figure workable:

In the final chorus of *Disorder at the Border,* a blues from the same session, he introduced another trademark: a flurry of quick notes, with some notes clearly articulated and others indistinct. In both solos he demonstrated his commonality with Charlie Parker, by using Figures 1A, 2A, 3, 4A, 4C, 4E, 5A, 6A, and 9 (pp. 31–33).

In January 1945, nearly eight years after making his recording debut, he had his first session as leader. The sextet for the date contained both swing and bebop players. They recorded *I Can't Get Started, Good Bait, Salt Peanuts,* and *Be-Bop* (see Ch. 2, p. 13). In *I Can't Get Started,* a ballad that had become a trumpeter's specialty ever since Bunny Berigan recorded it in 1936, Gillespie made two lasting contributions to the bebop language. First, he reharmonized the third and fourth measures of the *a* section, substituting four pairs of chromatically descending ii-V chords for the original, simpler chords.[7] Second, he created a coda that later became the standard introduction for Monk's *'Round Midnight.*

In Chapter 2, I referred to his fine melody *Groovin' High,* and to two recordings he made of it in 1945. There is a third one he and Parker did at the end of the year. The two players had taken a band from New York City to Los Angeles, and during their initial stay on the West Coast they performed this composition on a program for the Armed Forces Radio Services. In this extended version Gillespie's solo is two choruses long, in D♭.[8] Near the beginning he plays a characteristic short-note phrase, but sloppily, as though he were more interested in the overall gesture made with such a phrase and less concerned with individual notes. Another characteristic phrase occurs at the midpoint of his first chorus: a short rise to a high note followed by a descent:

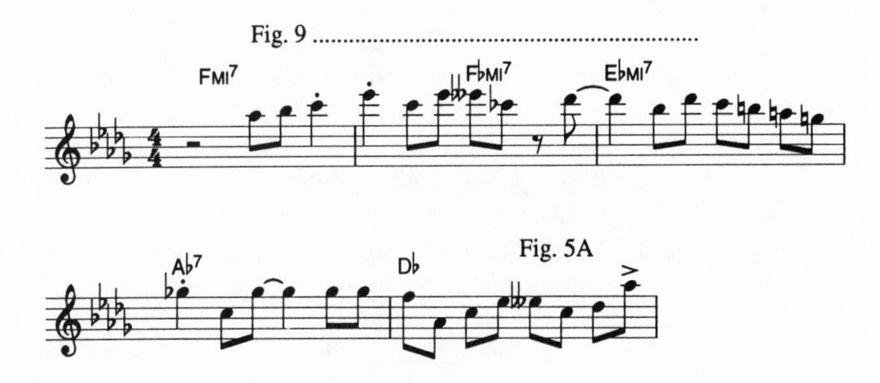

This particular descent he executes much as Parker did it—by using a decorated scalar descent from the high E♭ to the penultimate D♭ a ninth lower. But in many other solos the short rise to the high note leads to a rough-and-tumble descent in short, somewhat blurred notes, which Gunther Schuller likens to "some incredible musical slalom" (Schuller 1989: 453).[9] Also present in the example above are Figures 9 and 5A (pp. 32–33). The latter involves scale degrees ♭9-7-8 once again; Gillespie used this melodic ornament over and over, particularly at phrase endings, as in this case. Elsewhere in this solo is an early and inexactly played example of an Eldridge-inspired phrase; it became a signature phrase for Gillespie, one that he later wrote into his big-band arrangement of *One Bass Hit II*:[10]

His chorus on *Confirmation* from early 1946 is one of his best recorded solos from the 1940s.[11] It is less flashy than usual for him; the highest note is only C# above the staff. He plays effectively here, using a cleanly articulated sixteenth-note phrase in the bridge which he integrates well with the following phrase. Throughout the chorus are melodic figures that Parker often played; they, along with the descending scalar structure of several of the phrases, illustrate once more the similarities between the two players' musical vocabularies.

In 1945 Gillespie organized a big band, but financial difficulties led shortly to its dissolution. That band did not record, but his second band did, starting in June of 1946 (see pp. 20–21). Because of the expanded ensemble, for which written-out passages are the norm, and the limited recording time available on the standard ten-inch 78-rpm record, most solos are short in big-band recordings.[12] As a result, most of Gillespie's best solos occur elsewhere. Perhaps his best big-band solo is in *Two Bass Hit*,[13] which also has some excellent bass playing by Ray Brown. In 1950 financial pressures again forced Gillespie to give up his big band and resume small-group performances.[14] From that point on he led big bands only for brief periods—most notably during 1956–58, but also for individual concerts and festival appearances. For most of his professional life he led small groups, especially quintets.

Because of his pioneering role in bebop, Gillespie was most influential during the 1940s. But his finest trumpet-playing years were the 1950s, when the attention of the jazz world was focused more on several younger players than on the older master. The list below includes a number of solos that are far better than any he recorded in the 1940s:

Relaxin' with Lee, takes 4 and 6 (*Stompin' at the Savoy*)—6 June 1950, Verve 8006—from a session led by Parker; Gillespie solos equally as well in D (take 4) as in D♭ (take 6)
Birks' Works (blues in minor)—13 January 1951, Oberon 5100—taken from radio broadcasts with his sextet, which featured John Coltrane
Anthropology (I Got Rhythm)—31 March 1951, Alamac 2430 and other bootleg issues—from a radio broadcast with Parker
Boopsie's Blues, take 1 (blues)—16 August 1951, DeeGee 4000
Star Dust (ballad)—25 October 1951, DeeGee 3607—his tone is unusually full-bodied in this piece
Birks' Works (blues in minor)—9 February 1953, Vogue 574-30—from a concert in Paris
Hot House (What Is This Thing Called Love?) and *A Night in Tunisia*—15 May 1953, Debut 4 and several bootleg LPs—from the Massey Hall concert with Parker
Caravan and *Tin Tin Deo*—13 March 1955, Elektra Musician 96-0300-1—from a concert with The Orchestra in Washington, D.C.
Dizzy Meets Sonny (I Got Rhythm in B♭)—12 January 1956, Norgran 1076
Wheatleigh Hall (blues), *Sumphin'* (blues), and *Con Alma*—11 December 1957, Verve 8260
The Eternal Triangle (I Got Rhythm, with a different bridge) and *After Hours* (slow blues)—19 December 1957, Verve 8262
St. Louis Blues (blues)—17 February 1959, Verve 8313
Woody 'n' You—20 February 1959, Verve 8313
In a Mellow Tone (Rose Room) and *Perdido*—April 1960, Verve 8386—from an album of Ellington tunes scored by Clare Fischer
Salt Peanuts (I Got Rhythm in F)—9 February 1961, Verve 8401—concert recording
Here 'Tis (blues)—24 July 1962, Philips 600-048—concert recording in France
Dizzy's Blues (=Birks' Works)—1 October 1967, Solid State 18034—from a jam session at the Village Vanguard
Tin Tin Deo—14 November 1971, Atlantic 2-905—from a concert in London by "The Giants of Jazz"; a duet with bassist Al McKibbon, in which Gillespie plays both trumpet and piano
Constantinople (blues in minor)—27 April 1976, Pablo 2310 781—from a recording session with Benny Carter

Many of these recordings contain much longer solos than occur in his earlier recordings. In them, several other figures, some of which appeared occasionally in the 1940s, emerge clearly as signature phrases. This first example usually appears at a phrase beginning in the middle of a solo; it follows a rest of several beats, while he prepares for this dramatic high-note entrance:

Another phrase opening is even more ear-catching because of its emphasis on the ♭9 and #9 of a dominant chord:

He used the diminished scale in several ways; this, his favorite configuration, he may have borrowed from Miles Davis (see below):

The next figure is from the *Hot House* theme, which he recorded with Charlie Parker in 1945:

In this connection, he also liked to quote the beginnings of popular songs, just as Parker, Gordon, and many other players did. Perhaps his favorite was *Rain on the Roof:*

And finally, there is this simple but rhythmically effective figure:

After the 1950s Gillespie's tone became thinner and slightly pinched. This tone quality seemed deliberately produced, for in at least half of his solos he used the Harmon mute with the stem removed; there could hardly be a more constricted trumpet sound. In addition, he usually played with little or no vibrato; the warm vibrato he used in *Brother K* from 1989 (A&M 6404) was a rarity for him. Yet his music is far from austere sounding. No trumpeter used half-valving more effectively while playing a slow blues. And the high-note, attention-grabbing phrases and the rapid flurries of notes remained very much a part of his idiom.

In his last years, even more than in the 1940s, Gillespie made an unforgettable picture when performing. His rubbery cheeks and neck would inflate to an astounding size. In addition, he played a uniquely shaped trumpet, one with the bell aimed upward at a 45° angle. This unusual design was the result of an accident. In 1953 someone fell on his trumpet, which was sitting upright on a trumpet stand, and bent the bell back. Gillespie played it, discovered that he liked the sound, and had trumpets built in that shape from then on.

This sight, of course, inspired both amazement and amusement, which was entirely consistent with the impish demeanor and quick wit with which he handled life. Yet once the novelty of his unusual appearance wore off and the laughter generated by his humorous remarks died away, one noticed the underlying seriousness with which he approached his art. His was the work of a jazz master, one who helped shape the tradition for half a century. His importance to the tradition is confirmed not only by his acceptance into the pantheon of jazz's greatest creators, but by the many trumpeters who utilize his musical thoughts. Indeed, except for those few who look strictly to earlier players for inspiration, there may be no jazz trumpeters born after 1920 who do not derive at least some part of their style from Gillespie's.[15]

Howard McGhee (1918–87) was a transitional figure. His recorded solos from the early and middle 1940s suggest that he listened to Gillespie's early recordings (or did both men draw from a third source?). Both players used the ♭9-7-8 figure, both sometimes played "swing with funny notes" (see pp. 8–9), both used similar high-note flourishes and sixteenth-note flurries, all of which pointed to the future. At the same time, McGhee's rapid and automatic vibrato, his mostly diatonic melodic vocabulary, and his on-the-beat, rather than off-the-beat, stressing of notes on either side of an inverted mordent—

rather than

—suggested swing, not bebop.

If McGhee's music was only partly bebop, his involvement with bebop players was extensive. In 1946–47 he did his best to bring some stability into Charlie Parker's chaotic life during Parker's traumatic first visit to Los Angeles. He was the trumpeter on an infamous Dial session when Parker, teetering on the brink of a nervous collapse, managed to record *Max Is Making Wax, Lover Man,* and *The Gypsy.*[16] Seven months later he was on Parker's *Relaxin' at Camarillo* date. During that period he was a central figure in the Los Angeles bebop world, taking part in numerous concerts, recording, and even running a night club for a time. On his own recordings he often played the new bebop tunes, sometimes even claiming other's pieces as his own. His *High Wind in Hollywood* is actually Thelonious Monk's *Fifty-second Street Theme,* his *Hot and Mellow* is actually Parker's *Yardbird Suite,* and his *Messin' with Fire* is actually *Donna Lee* by either Parker or Miles Davis (sources do not agree).[17]

As time went on McGhee adjusted his style to some extent. Albums such as *Together Again!* with Teddy Edwards show that he could produce capable bebop.[18] Unfortunately, in later years his playing became sloppy; he often executed sixteenth-note flourishes with few recognizable pitches, and often fluffed notes in eighth-note phrases. His role throughout his career was that of a second-string player.

Joe Guy (1920-*ca.* 1962) was another transitional figure in the formative years. He led the house band at Minton's in 1941, and some of his solos are on the amateur recordings that survive from that location. He, along with Gillespie and McGhee, liked the ♭9-7-8 figure that they all inherited from Eldridge. And in a solo from 1942 he used a chromatic triplet-eighth-note figure in B♭ that perhaps is the source of Gillespie's signature figure shown earlier.[19]

Theodore "Fats" Navarro (1923–50) was a highly musical bebop trumpeter. During the late 1940s he and Gillespie were the most accomplished trumpeters in bebop. He drew extensively upon Gillespie's melodic vocabulary, using Gillespie's sixteenth-note signature phrase from *One Bass Hit,* his upper-register rise in quarter notes (illustrated above), and other phrases. But he was a less dramatic player than Gillespie, for he used the upper register less frequently and played fewer fast-note passages. Also, compared with Gillespie, he had a cleaner technique, a sweeter tone, and a more expressive vibrato. He had the most lyrical trumpet style in early bebop. Sadly, his important recordings span only six years—from 1945, when he replaced Gillespie in Billy Eckstine's big band, until shortly before his death in 1950. And for nearly all of his recordings he was a sideman on someone else's session (Kenny Clarke, Coleman Hawkins, Bud Powell, Tadd Dameron, and others), and few listeners outside the bebop cognoscenti knew his music during his lifetime.

Navarro borrowed melodic figures and some habits of phrasing from Parker as well as from Gillespie, and the architecture of his best solos is close to that of Parker. But, as with all other first-class improvisers, his style is not a carbon copy of any one style; he took elements from the vocabularies of his predecessors and mixed them with elements of his own to create a distinctive style. For example, in *Everything's Cool,* he played a phrase that begins with Figure 4E (see p. 32), embellished with the inverted mordent, 2A.[20] But the next phrase is more Navarro than Parker—

—for he liked to end phrases on scale degree 6 and warm that note with a gentle vibrato. (Clifford Brown also used this ending from time to time in the 1950s.) A longer passage in *Wail,* take 3 (cited in the list below), shows a similar mix of ingredients.

It contains a liberal sprinkling of Parker figures, as shown. The nearly continuous flow of eighth notes garnished with occasional accents between the beats is also a Parker trait. The effective repetitions of Figure 4D, however, which connects by chromatic descent the thirteenth and augmented eleventh (E and C# in the first two measures, D and B in the sixth measure, G and E in the seventh to eighth measures) of three different chords, is a Navarro touch.

Among his best solos are the following:

Our Delight, 2 takes—26 September 1947, Blue Note 540 and 1531—with the Tadd Dameron Sextet

Fats' Flats and *Koko*—8 November 1947, Spotlite 108—with Barry Ulanov and His All-Star Metronome Jazzmen on a radio broadcast with Parker and others

Nostalgia and *Barry's Bop,* 2 takes each—5 December 1947, Savoy 955 and 2216

Wail, takes 2 & 3—8 August 1949, Blue Note 1531 and 1567—with Bud Powell's Quintet

Rifftide, Cool Blues, and *Ornithology*—30 June 1950(?), Charlie Parker 701—from a radio broadcast from Birdland with the Charlie Parker Quintet

The last three pieces in this list are particularly intriguing. The date given by discographers is only a week before Navarro died of tuberculosis exacerbated by heroin addiction. Surely he was emaciated and gravely ill, yet he played with undimmed skill and inventiveness. If that date is accurate, these recordings represent an amazing triumph over physical adversity.

Other bebop trumpeters in the 1940s helped establish and spread the new bebop vocabulary for their instrument, though they were less innovative than Gillespie, Navarro, and Davis. Among them were Kenny Dorham (1924–72) and Red Rodney (1927–94).

Dorham played in the big bands of Billy Eckstine and Dizzy Gillespie in 1946, and beginning in that year also played on several important small group recording sessions. In December 1948 he replaced Miles Davis in Charlie Parker's quintet. But then his technique was imprecise, his tone was nondescript, and his melodic ideas were largely derivative of better players. He made his best recordings in the 1950s; well-crafted, expressive solos played with an improved tone quality include *Let's Cool One* with Thelonious Monk, *Hippy* with Horace Silver, *Soft Winds* with Art Blakey, *Ezz-Thetic* with Max Roach, and his own *Blues Elegante*.[21] During the 1950s he may have contributed the extended zig-zagging chromatic descent that Miles Davis used extensively in the 1960s (p. 126). Also, he certainly kept this figure alive, by using it repeatedly in almost every solo:

He made perhaps his most important contributions to the language with his graceful compositions *Blue Bossa* and *Prince Albert* (the latter, an *All the Things You Are* contrafact, co-written with Max Roach).

Red Rodney also had an early career in big bands (with Claude Thornhill and Woody Herman, among others), and in late 1949, he replaced Dorham in Parker's quintet. Like Dorham in those early years, Rodney was a competent but unexceptional player. In the mid-1950s he left the jazz world to become a society band leader; from there he became a securities expert, a bank bilker, a prisoner, a law student, a Las Vegas show musician, and—in the 1970s—a jazz musician once more.[22] The turning point came in 1979, with the formation of a quintet co-led at first by multi-wind instrumentalist Ira Sullivan. His career centered on the quintet format ever since then. Within this stimulating musical context he played bebop with great naturalness, conviction, and expressiveness, particularly on flügelhorn. The former young journeyman became a sexagenarian master.

Clark Terry (born in 1920) belongs in a survey of bebop players, although he built his reputation as a sideman in the swing-style bands of Charlie Barnet (1947), Count Basie (1948–51), and Duke Ellington (1951–59). His early solos reveal the influence of Dizzy Gillespie, primarily in high-note passages. Yet his distinctive way of lipping up to notes, his meticulous and lengthy sixteenth-note passages, and his mellow tone quality set him clearly apart from Gillespie. In the late 1950s Terry began doubling on flügelhorn; eventually it became his primary instrument. On this larger instrument his already rich tone is even more mellow, and his style is among the most instantly recognizable in jazz.

Terry has one of the most contagious senses of humor in jazz. It surfaces in several ways. First is his use of theme quotes, a standard practice among many beboppers. Second is his scat singing. He and Gillespie are two of the greatest scat singers in jazz. But his repertory of vocables is unlike Gillespie's, or anyone else's; he regularly adds English words to the mix and uses the inflections of heightened, emotional speech. These almost, but not quite, intelligible "lyrics," which he calls "mumbles," are often quite funny. Third is his remarkable alternation between trumpet (usually with a Harmon mute) and flügelhorn during a solo. He sometimes plays a chorus on trumpet followed by a chorus on flügelhorn; then, holding one instrument in each hand, he switches back and forth between the two in ever shorter spans of time. The unsuspecting listener to his recordings may wonder how two players can jointly improvise a melody so telepathically, not realizing that it is Terry alone. But anyone watching him do this feat would laugh—at the unusual sight, and in appreciation of his technical expertise.

His bebop lines, laced with slyly humorous quotations and playful turns of phrase, stand out on numerous Ellington recordings of the 1950s. His best-known solos were on *Perdido,* which Ellington assigned to him as a feature number.[23] After leaving Ellington his associations with big bands continued. For years he played in big bands for NBC, most notably the *Tonight Show.* He also made valuable contributions to big-band albums led by Quincy Jones, Gerry Mulligan, and others.[24] From time to time he also has led his own big band and various small groups.

Terry's sixteenth-note passages frequently contain multiple repetitions of short figures played with clean and crisp articulation. The two figures shown here are typical:

Much of his unique sound stems from his tone quality, phrasing, and articulation, for his sound comes through clearly even in written-out ensemble passages. A wonderful sampling of his phrasing and articulation habits occurs in *Tête a Tête*,[25] based on the chord structure of *Honeysuckle Rose*. The lipped notes in measure 18 almost sound like someone saying "er-ow":

In this example the hallmarks of his style include the portamento of measure 18, the lack of vibrato where one would expect it (the long notes of measures 21 and 25), and the unexpected vibrato in the middle of the phrase (measure 23). This excerpt also contains an elegant prolongation of the structural tone C (scale-degree 5 in F major), accomplished with the ornamented scalar descent that Parker favored, but done largely without Parker's figures (see the graph—line 4).

Terry's 1964 recording of *Star Dust* (Emerald 1002), which became available only in 1987, is an extraordinary sample of Terry's unique flügelhorn tone quality. It is not a uniformly fine performance, in part, perhaps, because the small audience sounds only mildly interested in the music. But Terry's masterful rendering of the *Star Dust* verse—unaccompanied at first and then joined by the bassist—is stunning. That moment is a gem; how fortunate we are that someone recorded it!

Typically jazz musicians find their basic playing styles during their early twenties, and spend their careers polishing and refining those styles. Miles Davis (1926–91) was hardly the typical jazz musician, however; for the first 20 years of his career his personal style continually evolved. Few other major players moved through more musical stages over such a long period.

His early contacts with the professional jazz world produced mixed impressions. While still in high school in St. Louis he received job offers from bandleaders Tiny Bradshaw, Earl Hines, Jimmy Lunceford, Illinois Jacquet, and others (which his mother made him turn down so that he could finish high school—see Carr 1982: 9–10). Yet when he filled in briefly for an ill trumpeter in the Billy Eckstine big band, Eckstine was unimpressed. "When I first heard Miles," Eckstine recalled, "I let him sit in so as not to hurt his feelings, but he sounded terrible; he couldn't play at all."[26] These events took place during World War II, when seasoned players were scarce in the civilian world, which may explain how a "terrible-sounding" player could have been offered jobs by major band leaders.

Three months after graduating from high school in 1944 he moved to New York City, officially to study at Juilliard School of Music, but in fact to immerse himself in the center of the bebop world. Soon he had renewed his friendship with Charlie Parker, a friendship begun when he had played with the Eckstine band in St. Louis, and was spending his nights listening to Parker and the other leading beboppers. A year later he began playing regularly with Parker in clubs on 52nd Street, thus beginning a professional association that would continue intermittently for several years. On 26 November 1945 he recorded his first solos on Parker's famous first recording date as leader (see pp. 17–19).

Davis's solos on various takes of three different pieces from that session show that at age 19 he had good intonation and a fine tone but an immature sense of melodic development. The warm, lyrical sound of his long notes shows that tone quality was always important to Davis, who was to revolutionize the concept of jazz trumpet playing.[27] But the incipient beauty of his tone is overshadowed by his pedestrian improvisations, which pale in comparison with Parker's. Among other things Davis overworks a simple oscillating figure—

—and overworks a phrase ending on the ninth of the tonic chord. In his solos on *Now's the Time,* takes 3 and 4, Dizzy Gillespie's supportive, interactive comping on piano is more interesting than Davis's solo lines. His best solo is in *Thriving from a Riff,* take 3, the second half of which, while heavily dependent on borrowed figures, is well-crafted early bebop. For this solo Davis used a cup mute, which he used often during the late 1940s (the famous unstemmed Harmon mute sound was still several years away).

During the 1940s Davis experimented and grew musically. At times he tried some Parker figures; he also borrowed from Howard McGhee, Dizzy Gillespie (including Gillespie's signature figure in chromatic triplets), and others. The various takes from Parker's recording sessions show signs of his struggles to master the bebop language. In *Ornithology,* take 1, he loses his way harmonically near the end of his chorus; in *Cheryl* and *Buzzy,* take 3, his chromatic upper-neighbor figures stick out like sore thumbs, much as Gillespie's did in the Minton's recordings;[28] in several solos he attempts some unconvincing Gillespie-like flurries of sixteenth notes, or plays up to six pairs of oscillating notes as though he were stuttering.

On the positive side, however, he was developing an ever more lyrical sound and a distinctive melodic vocabulary. His sound quality seemed to leap forward in 1947, especially in his first session as leader, on 14 August. In *Milestones, Little Willie Leaps, Half Nelson,* and *Sippin' at Bells,*[29] his articulation of notes is smooth and gentle, his vibrato is even warmer than before, and his melodic lines are graceful and convincing. Six weeks later, in *Bongo Bop,* take 1, he showed that he could play cleanly executed sixteenth-note phrases of his own design. In measures 7–8 of his solo in *Out of Nowhere,* take 3, he plays a diminished scale figure that he was apparently the first to record (see p. 106 above). To compare his work in these solos, and especially in *Drifting on a Reed,* take 5, with his first solos recorded two years earlier is to hear a dramatically improved musician. Clearly by late 1947 Davis had become an important bebop player. Evidence of further growth appears in the broadcast performances of 11 December 1948, in which his lyrical, sweet sound on long notes commands attention. Along with the lyricism is an intensity, especially in *Ornithology,* that showed that there was more to Davis's style than just pretty notes. Although he uses his upper range sparingly, he hits high D confidently in *Ornithology* and *Big Foot*; in the latter piece he even hits a squeaky high F.[30] And in both solos he and Max Roach play a rhythmic game that they often engaged in during that time: building a string of short phrases on the same rhythm (one that Bud Powell also used in the melody of his *Dance of the Infidels*)— .

During 1948, while still working regularly with Parker, Davis joined a nine-piece rehearsal band, formed to play new arrangements by Gil Evans, Gerry Mulligan, and John Lewis (see pp. 23–24). When the group recorded, Davis had become the leader. Since their repertory favored slow or moderate tempos and quiet dynamics, many jazz writers consider this band as the beginning of the "cool era." Davis's short solos are hardly his best of the time. But his beautiful, lyrical sound prevails throughout these pieces. Also, he uses an important embellishment, a kind of "measured grace note," which became a stylistic element that many imitators copied:[31]

The stylistic growth heard in his recordings of the 1940s continued in the 1950s. He used some of Clark Terry's vibrato and tone quality in *Bemsha Swing* and other performances.[32] A striking new feature was his extensive use of separately articulated, evenly spaced eighth notes; in some solos swing eighths disappear entirely. Another new trait was an occasional phrase that begins on a high-note explosion and quickly descends; the explosion typically begins around high E♭ or F, but drops away so quickly that the starting pitch is unidentifiable. A more general feature is his emerging skill in ballad playing. During his sideman days his contributions to ballads were mostly brief and superficial, though pretty. But as a leader—after 1951 he rarely recorded as a sideman—his ballad recordings showed a greater maturity. His first important ballad recording is *My Old Flame,*[33] a two-and-one-half-chorus showcase of his lyrical style. It dates from the first session he led in which he plays lengthy solos. This "blowing session"—the first of many such sessions in the 1950s and 1960s—contains especially fine solos on *Dig* and *Bluing,*[34] and provides the first traces of the enormously important "sound" that he would develop fully by 1955.

Unfortunately, Davis's jam-session approach to recording sometimes was too casual. Apparently he waited until the record date began to pick the songs, and if no one in the group knew a chosen song completely, Davis (or the group) would make up the unknown portions. The most infamous example of recomposition is the bridge of Benny Carter's jazz standard *When Lights Are Low.*[35] Rather than using Carter's nicely crafted tour through the circle of fifths, Davis and his colleagues John Lewis and Percy Heath simply used the melody and chords of the *a* section transposed up a fourth. This solution is workable and logical, but is much less inventive than Carter's, and, of course, is blatantly incorrect. The recording became well known and widely copied, both on records and in public performances.[36] Another case is Davis's 1954 recording of Thelonious Monk's *Well, You Needn't* (Blue Note 5040). In this case Davis and his colleagues Horace Silver and Percy Heath are closer to the original composition, but they alter the main motive of the *a* section, start the bridge melody a half-step too low, and use a reharmonization for the bridge that is incompatible with Monk's harmonization.[37]

Davis's recordings made in 1954 contain several new elements that point toward that world-renowned style of the late 1950s. Here and there, especially when the chord changes occur frequently or are unusual, he constructed chromatic melodies that clash with those chord changes, sometimes for several measures at a time. Also he hinted at the poignant expressiveness with which he imbued long notes later. And on 3 April 1954 he recorded using the Harmon mute for the first time. The artistic high points of the year for him, however, are two open-horn blues played with relatively little of the late-1950s "sound": *Walkin'* and *Blue 'n' Boogie* (Prestige 1357 and 1358). In these pieces he received great support from Silver, Heath, and Kenny Clarke; together the group produced some of the year's best bebop.

The year 1955 was a watershed for Davis. In June he made his first recording session with pianist Red Garland and drummer Philly Joe Jones. In the fall he organized a quintet with these two players plus John Coltrane and bassist Paul Chambers, forming what soon became a legendary bebop ensemble. Equally important, in the June recording session and in later sessions, he documented his famous "sound."

The Miles Davis "sound" contains more than his notes, phrasings, and articulations, although it included those elements. The principal components in 1955 were these:

1. A simple, straightforward performance scheme that occurs often in unrehearsed "blowing sessions": piano introduction, theme paraphrase by Davis, solo choruses by Davis and others, closing theme paraphrase by Davis. The only organizational detail that might have required much discussion is the use of codettas—repeated four-measure turnarounds at the end of each solo (used for the first time on *'Sposin'*[38]).

2. A casual approach to the theme. He frequently played skeletal versions of the theme with many rhythmic displacements of individual notes; no one could have played a unison line with him.

3. The frequent use of the Harmon mute, always without the stem.

4. A remote-sounding presence, with a modest amount of echo, for the open-horn solos. Recording engineer Rudy Van Gelder first created this effect in the November recording session.

5. Simple improvised melodies, often harmonically non-specific, played with impeccable swing.

6. An occasional chromatic phrase having little or no obvious connection with the harmonic structure. The bridge sections of several

choruses in *I Didn't,* from the June recording session, contain clear examples of freely chromatic, "outside" phrases.[39]

7. An occasional low note played without vibrato.

8. A characteristic way of playing lower-neighbor grace notes (shown in the last musical example).

9. A quick crescendo and upward slur to pull from one note to the next. The crescendo results from a muted-to-open articulation, as in this example from *Will You Still Be Mine:*[40]

This articulation pattern—a kind of note "squeezing"—is particularly striking on higher notes played with the Harmon mute, as in this example from *A Gal in Calico:*

10. A terminal vibrato pushing the end of a note just to the beginning of the next beat (see the example from *Will You Still Be Mine*). This device was common in jazz ever since Louis Armstrong perfected it in the 1920s, but Davis's gentle, subtle vibrato and his gorgeous tone create a whole different effect.

11. Rapid repeated-note figures, produced by alternate fingerings or by tonguing.

12. A cascading scalar decent, as in this example from *There Is No Greater Love:*[41]

During 1955–60, Davis applied his sound to a repertoire based on blues, other jazz standards, and popular standards. The popular standards—generally played in slow and moderate tempos with an unstemmed Harmon mute—attracted the most attention.[42] Many writers thought of these subdued performances as a culmination of the cool era in jazz that began with his nonet recordings of 1949–50. But despite his sweet tone, lower dynamics, and slow tempos, the essential rhythmic, harmonic, and melodic vocabularies that he and his sidemen used were clearly from the bebop idiom. Also, his fluent handling of fast tempos in *Airegin, Dr. Jackle, Walkin'*, and others showed that his playing could fit anyone's definition of bebop.[43] Miles Davis's sound in the 1950s did not represent a separate general jazz style, but rather an expansion of the parent style.

In 1956 he added a short, poignant fall-off to his sound, and by the following year it was one of his most expressive effects. It is the opposite of the muted-to-open effect discussed above, for it includes a brief descrescendo, as he drops his jaw and air support:[44]

A second late 1950s trait is so simple it needs no notated illustration—a long, low note warmed with vibrato. Words fail utterly to express the beauty of those occasional Fs and Es; the rich, full tone, the quiet dynamic, the subtle vibrato all combine magnificently. These two effects expanded his arsenal of articulation devices, and he used them often to work musical magic with the simplest melodic gestures. Some of Davis's best recordings from the late 1950s and early 1960s are these:

Bye, Bye, Blackbird—5 June 1956, Columbia 949
All of You—10 September 1956, Columbia 949
If I Were a Bell—26 October 1956, Prestige 7129
Blues by Five (blues)—same date; Prestige 7094
Miles Ahead (entire album)—May 1957, Columbia 1041
Four and *Bye, Bye, Blackbird*—13 July 1957, Jazzbird 2005—from a jazz-club broadcast performance
One for Daddy-o (blues in minor)—9 March 1958, Blue Note 1595—from a session led by Cannonball Adderley
Straight, No Chaser (blues) and *Miles* (called *Milestones* on later performances, but unrelated to the *Milestones* of 1947)—2 April 1958, Columbia 1193
Love for Sale—26 May 1958, Columbia 33402
Summertime—18 August 1958, Columbia 1247—from the album *Porgy and Bess,* with an ensemble arranged and led by Gil Evans

Kind of Blue (entire album)—March and April 1959, Columbia 8163
Solea—11 March 1960, Columbia 8271—from the album *Sketches of Spain,* with an ensemble arranged and led by Gil Evans
No Blues (a.k.a. *Pfrancing;* blues)—13 October 1960, Dragon 129/130—from a concert in Stockholm
No Blues—19 May 1961, Columbia 8612—from a concert in Carnegie Hall
Seven Steps to Heaven—14 May 1963, Columbia 8851
So What—12 February 1964, Columbia 9253

There are two landmark albums on this list: *Miles Ahead* and *Kind of Blue.* The first places Davis in a large-ensemble setting devised by Gil Evans. Throughout the album Davis plays flügelhorn, an instrument he rarely played. Most of the pieces are slow and lyrical, and Davis responds appropriately with his beautiful sound, made even more mellow here because of the flügelhorn. The beauty of his sound in pieces such as *The Meaning of the Blues* and *Lament* is extraordinary. The second album, recorded two years later, consists of five pieces, each a superlative model of mid-twentieth-century jazz.[45] Davis's solos in these pieces are excellent for different reasons, including the poignant beauty of his muted solos in *Blue in Green* and *Flamenco Sketches,* the understated but relentless swing of the blues solos *Freddie Freeloader* and *All Blues,* and the motivically unified exploration of the Dorian scale in *So What.*

Blue in Green, during Davis's two solos, is a ten-measure ballad (during choruses by Coltrane and pianist Bill Evans the structure changes—see pp. 160–61). His elastic rhythms, variety of attacks and releases, and sparing use of vibrato are all typical of his 1950s approach to ballads. Yet something—perhaps the lyrical quality of the theme, perhaps the exquisite accompaniment provided by Evans, Chambers, and Cobb—inspired him to play in a particularly effective way here. The Harmon mute and the frequent long notes without vibrato create an austere effect as he paraphrases the gentle theme. There is nothing pretty in these four choruses, but there is much that is beautiful. His two solos in the flexibly structured *Flamenco Sketches* have much the same general sound; the two pieces have nearly the same tempo, and again he uses the Harmon mute and almost no vibrato. But *Flamenco Sketches* has only a slow-moving harmonic scheme, no precomposed melody. Thus, since there is nothing to paraphrase, Davis's two solos are almost totally different from one another melodically.

The theme of *All Blues* has two main motives, an ascending sixth and a neighbor-tone embellishment of the higher note:

During his two opening and two closing theme choruses Davis plays these connected motives in several different rhythmic configurations. When he removes his Harmon mute and plays his four-chorus solo he uses ascending leaps and neighboring motions (both upper and lower) extensively:

More prominent, however, is a simple two-note rhythm—the well-known "be-bop" rhythm ( ). It occurs 21 times, mostly on two repeated Gs or on the descending third B-G. Thus, he unifies his solo beautifully. And as usual, he applies his extensive array of performance techniques—long notes without vibrato sometimes, subtle vibrato at other times, bent notes, "squeezed" notes, fall-offs, and so on—to make this a masterful, classic solo.

By far the most famous Davis solo in *Kind of Blue* is in *So What.* Perhaps thousands of jazz students have learned this solo, or portions of it, note for note, for it is a wonderfully simple and effective illustration of how to build a solo on a stationary harmonic structure. Writers have used the terms "modal jazz" and "scalar jazz" to describe *So What* and similarly constructed pieces, since the theme derives almost entirely from the Dorian scale on D and on E♭.[46] But the musicians playing such pieces regularly depart from a strict usage of one particular scale during their solos. Davis, for example, uses an occasional C# as a lower neighbor to D, a G# as a lower neighbor to A, and an A♭ (the "flatted fifth" in D) as an upper neighbor to G. And since the accompaniments usually stay centered on a tonic chord, as in an ostinato or "vamp," Barry Kernfeld suggests that this solo illustrates Davis's "vamp style" (Kernfeld 1981: 128–74).

Davis's solo is a model of economy; he plays only 3.4 notes per measure on the average (in contrast, Coltrane plays 5.8 notes per measure in his solo in the same recording). Fully a third of his 64-measure solo has zero, one, or two notes per measure; musical space is obviously important to him. After a three-note lead-in Davis begins with a figure that becomes a unifying motive:

This motive dominates the first eight measures, and variants of it reappear in the second and fourth eight-measure sections and in the final measures of the solo. A sub-motive here is the final pair of notes—the "be-bop" motive again. This syncopated phrase ending recurs five more times. And a simple unfolding of minor-eleventh chord tones (shown in whole notes below) becomes a third motive.

Touched on fleetingly in the first chorus, the motive dominates the first eight measures of the second chorus, and appears prominently in the last sixteen measures of the solo. Throughout the solo his uniquely beautiful tone, unerring timing, and varied attacks and releases serve to bring his simple melodic ideas to life:

A month after recording this historic version, Davis performed *So What* on a television show.[47] In this version Davis played two two-chorus solos that provide a fascinating glimpse into his creative processes. His notes-per-measure ratio is about the same as in the original solo, and his favored phrase ending is again the "be-bop" figure, which appears 13 times in these two solos. Also reappearing from the original recording are the minor-eleventh chord unfolding and a nearly identical vocabulary of pitches. But two different figures play large roles in the April 2 performance. The first involved the "flatted fifth"; it appeared only once in March (see the last measure in the example above), but nine times in April. The second is a scale passage in D Dorian that starts on scale degree 1 or 2 and ascends an octave or more in eighth notes. Still, the overall effect of all three pairs of choruses is similar. The big changes in his approach to *So What* were yet to come.

By the end of the decade, Davis's tendency in improvising was to de-emphasize the harmonies of themes and to focus instead on rhythm, melody, and articulation. In so doing he was harking back to the standard approach to improvisation taken by Louis Armstrong, Lester Young, and many other pre-bebop players, although his style otherwise was far removed from those older styles. Surely he knew the chords; in many passages his melodies are as harmonically explicit as Charlie Parker's. More often, however, he avoided arpeggiating chords or emphasizing the harmonically explicit chordal thirds and sevenths, and instead favored harmonically neutral notes. For example, his first solo chorus in a concert performance of *Walkin'*,[48] contains only the first four notes of the blues scale (F-A♭-B♭-C♭):[49]

He had a special vocabulary reserved for a small body of pieces, those with a Spanish tinge. Most of these pieces contained sections based on the "Spanish Phrygian" scale—the Phrygian scale with the minor and major thirds used interchangeably. The first such piece in his repertory was *Flamenco Sketches.* In one portion of this piece Davis built his phrases on the D Phrygian scale (D-E♭-F-G-A-B♭-C-D) while Bill Evans based his accompaniment on the D major triad, which contains an F#. In almost every phrase Davis plays the distinctively Phrygian notes D-E♭-F. Consequently, these melodies have a different quality from the major-minor-based main part of his repertory. In later pieces he went further toward developing his Spanish style, by employing some "crying" embellishments that are rare in jazz. The introduction of *Solea* from the album *Sketches of Spain* contains a stunning example:

free meter

(lip)

This small but fascinating body of Spanish-tinged pieces includes everything else in the *Sketches of Spain* album, plus *Teo, Mood, Masqualero, Spanish Key,* and *Fat Time.*[50]

In the 1960s Davis continued to use repertory from the 1950s in concert performances; short versions of *The Theme,* plus *So What, Walkin'* and *Pfrancing* (both blues in F), *All of You, If I Were a Bell, 'Round Midnight,* and several others turned up repeatedly on concert performances. But the approach taken to some of these pieces was much different from the relaxed, "cool" approach of earlier years. The tempos of *So What, Walkin', Miles* (or *Milestones*), and others are often ♩ = 300 or faster. And Davis's playing style is much more aggressive than before; he is louder, plays more notes per measure, and plays more high notes than he did earlier. Part of this style change might be attributed to his new sidemen—by 1963 his dynamic rhythm section was pianist Herbie Hancock, bassist Ron Carter, and drummer Tony Williams—but the stylistic shift started in 1960, before these men joined his group.

With the new aggressive approach to improvising came additions to his vocabulary. One new pattern is a simple string of repeated eighth-note pitches, played with alternating fingerings. Several of his solos on *So What,* for example, begin with repeated Ds. Almost as simple is a rapid, unmeasured oscillation, usually around the note A:

And surprisingly, he returns to the oscillating figure that plagued his earliest solos in 1945; now, however, he integrates it convincingly into the total phrase. Another pattern is really two melodies in one, a contrapuntal melody. However, the lower notes of the pairs are so soft they are all but inaudible, and their specific pitches are probably incidental byproducts of his opening and closing of the air stream with his tongue ("da-n da-n da-n . . ."):[51]

In *So What* he likes to embellish a scalar descent as follows—

—and to interpolate a quote from the old popular song *I Found a New Baby*—

This last example is only a small sample of his often overlooked musical puns. From his first recordings in 1945—when he worked *Singing in the Rain* and *Thanks for the Memory* into his solos—to the end of his life, he played this standard bebop game. His young sidemen in the 1980s and 1990s may not have recognized these musical jokes, but he still played them, nonetheless.

Davis's improvised lines in the 1960s contain much non-harmonic chromatic motion. For example, he often plays a chromatic ascent half an octave or more up to E♭. The E♭ may or may not be a chord tone; its harmonic significance seems unimportant to Davis. Another pattern is a zig-zag chromatic descent, an extension of a two-beat figure that he used occasionally in the 1950s, and which he may have learned from Kenny Dorham:

Again, the starting and ending notes of the phrase and the notes falling on downbeats may not have any harmonic significance; the phrase is simply a means of moving through time. He relies entirely on the pianist and bassist to maintain the tonality and clarify the harmony in such moments.

In the early 1960s Davis and his quintet maintained the 1950s repertory in public, but recorded a radically different repertory in studio recording sessions, one that moved to the very edge of bebop and finally went beyond it. The trend away from bebop began during a session in May 1963, the first in which the 36-year-old Davis had as his sidemen tenor saxophonist George Coleman (age 27), Herbie Hancock (age 22), Ron Carter (age 26), and Tony Williams (age 17!). The most adventurous piece they recorded was Victor Feldman's composition *Joshua* (Columbia 8851), which has a main section in common time, but a bridge that is partially in triple meter. They also performed this piece publicly, and the concert performances pose a challenge to the listener, for the rhythm section does little to clarify the metric structure for the listener. Also, Davis, whose improvisations in the 1960s had become ever more chromatic, seldom articulates the harmonies.

The move away from mainstream bebop intensified in January 1965 (Columbia 9150). By then Wayne Shorter had become the quintet's saxophonist, and he brought to the recording studio *E.S.P.* and *Iris,* pieces with non-traditional harmonies. At the same time Ron Carter contributed *R.J.* and Davis contributed *Agitation,* both similarly adventurous pieces. In October they recorded *Orbits* and *Dolores,* angular melodies with obscure harmonic structures. With these pieces the group approached the non-tonal, non-chordal jazz of Ornette Coleman.

Some of their new repertory was less adventurous. In May 1963, when they recorded *Joshua,* they also recorded *Seven Steps to Heaven,* which has become a bebop standard. In 1965 and 1966 they recorded *Eighty-One* (Columbia 9150) and *Footprints* (Columbia 9401), which are blues in F and C minor respectively. However, the rhythm section, both in these simple pieces and in the boundary-stretching pieces, continually challenged the norms of bebop. Hancock's harmonies extended beyond the dissonances of the 1940s and 1950s; Carter often departed from the rhythmically predictable walking bass lines; and Williams often seemed willing to play anything except the normal bebop drum patterns. On top of this challenging accompaniment were the non-harmonic "outside" melodies improvised by Davis and Shorter. Consequently, even these relatively conservative pieces are far removed from the music Davis once played with Charlie Parker.

Davis's style changed little in his last 25 years, except to decline in quality for a time in the late 1970s and early 1980s, when he was seriously ill. But the background for his solos changed dramatically. In the late 1960s he began playing pieces with a rock-influenced accompaniment, complete with electric keyboards, electronically distorted guitars, ostinato bass lines, and rock-style drum patterns; thus, he entered the world of fusion jazz. However, he returned occasionally to bebop, as in *Kix, Mr. Pastorius,* and the soundtrack for the movie *Dingo.*[52]

Ever since the 1950s Davis played two conspicuous roles in jazz evolution. First, he was one of the most imitated players in the tradition. Imitation began in the late 1940s, when Shorty Rogers and a few others saw his music as an alternative to the technically demanding path cut by Gillespie. Later, Jack Sheldon, cornetist Webster Young, and others found his early to middle 1950s style appealing. Then Freddie Hubbard, Palle Mikkelborg, Dusko Goykovich, cornetist Nat Adderley, and others used components of his classic 1950s style. In the early 1980s, Wynton Marsalis dipped into Davis's stylistic bag of the 1960s. Second, because of his prominence his sidemen often became prominent as well. Cannonball Adderley, John Coltrane, George Coleman, Wayne Shorter, Steve Grossman, Gary Bartz, Dave Liebman, the two Bill Evanses, Red Garland, Wynton Kelly, Herbie Hancock, Chick Corea, Keith Jarrett, John McLaughlin, Ron Carter, Dave Holland, Philly Jo Jones, Jimmy Cobb, Tony Williams, and others probably would have made their marks in jazz even if they had not been in his groups. Still, they all gained critical international exposure while working for Miles Davis, which indicates both Davis's good musical judgment and his enormous impact on jazz history.[53]

In the 1950s a host of fine young bebop trumpeters expanded the bebop language by building upon the foundation that Gillespie, Davis, Navarro, and others established earlier. For a time there appeared to be two main stylistic paths to follow: the one established by the energetic, outgoing Gillespie and Navarro, or the more subdued one taken by Davis.

Chet Baker (1929–88), one of those inspired by Davis, garnered attention with his soft, gentle, narrow-range solos, first with the Gerry Mulligan Quartet and then in his own groups. He had a clean, precise technique and a flair for improvising graceful melodies. Unfortunately, he also had an anemic tone quality and a limited sense of drama. His typical solos are nice but bland. Eventually he expanded his emotional range somewhat, but a chaotic personal life dominated by heroin-related physical and legal problems kept him from achieving his musical potential.

In time the low-keyed approach lost favor among trumpeters. Even Miles Davis, the "founder" of "cool" jazz, moved on to other things, as we have seen. The more aggressive, emotional players, never in the minority, had completely overshadowed the quieter players in the second half of the 1950s, and have held sway ever since.

Clifford Brown, nicknamed "Brownie" (1930–56), was the most influential of these more dramatic younger players, and one of the greatest soloists in jazz. Before making his first recordings Brown had favorably impressed Parker, Gillespie, Navarro, and others with whom he had played. Yet his first two recorded solos, made early in 1952 with Chris Powell, only hinted of the greatness to come.[54] During the next fifteen months much musical growth took place. His second set of recorded solos, made in June 1953, reveal a distinctive new trumpet style, one that changed little during the rest of his short life.

During the first two years of his recording career Brown functioned primarily as a sideman. After the early session with Chris Powell he played sessions in New York City with Lou Donaldson, Tadd Dameron, and J. J. Johnson. He spent the fall of 1953 in Europe as a member of Lionel Hampton's big band. While there he did several small-group recording sessions, most with fellow Hampton-band member Gigi Gryce. When he returned to the U.S. he played briefly with Art Blakey. Fortunately a performance of Blakey's quintet performing at Birdland exists on records (Blue Note 5037-39 and LA 473 J2). It is Brown's first great recording, partly because he is in the stimulating atmosphere of a jazz club, partly because the rhythm team of Blakey, Horace Silver, and Curley Russell give him magnificent support.

In the spring of 1954 he and Max Roach began an immensely productive two-year partnership as co-leaders of a quintet. They formed the group in the Los Angeles area, using tenor saxophonist Teddy Edwards and pianist Carl Perkins at first. Soon the personnel was tenor saxophonist Harold Land (until December 1955), Sonny Rollins (Land's replacement), pianist Richie Powell (Bud Powell's younger brother), and bassist George Morrow, a group that made some of the finest bebop ensemble recordings of the decade (see pp. 216–18).

Brown's beautiful style of playing has been the envy of many trumpet players over the years, many of whom have striven to emulate him. The components of that style include:

1. A careful, even meticulous, adherence to the harmonies of most tunes. A wonderful example is his deft skating through the many chord

changes of *Joy Spring* (the first version, in E♭).[55] Yet he never simply "runs the changes." His melody is largely stepwise and he unerringly steps into harmonically significant notes at exactly the right moment. Perhaps his harmonic skill stems from his piano-playing abilities, which, according to Dizzy Gillespie, were considerable.[56]

2. A melodic vocabulary closer to Parker's than to Gillespie's. Parker's favorite figures were also Brown's, especially those labeled 1A, 1B, 2A, 3, 4A, 4B, 4E, 4F, 5A, 5B, 9, and 11A in Chapter 3. Further, his phrases often elaborated descending scales, although less consistently than Parker's melodies. But Brown could not help but be influenced by Gillespie—what young trumpeter could in the 1950s? His more ear-catching borrowings include Gillespie's sixteenth-note phrase from *One Bass Hit II* and his high-note, fanfare-like phrase beginning (see Chapter 6). He also used Gillespie's reharmonization of *I Can't Get Started.* His occasional song quotations are not attributable to any one player, however, since nearly everyone did it.

3. A highly refined sense of melodic development. He often filled his solos with long phrases, spinning them out effortlessly with an elegant balance between repetition and contrast. For example, in his final solo chorus of his earliest recording of *Daahoud*[57] he ended the first 8 measures with a chain of three motives, each of which he reused in the next 8 measures.

The syncopated motive *b* is clearly the most ear-catching because he used it so extensively, but the altered repetitions of motive *a* and motive *c* also add coherence to the phrase.

4. A preference for even eighth notes, separately articulated. Miles Davis's solos from the early 1950s were the probable source of this trait, but sometimes Brown played notes shorter than Davis did. At the start of one of his 1956 recordings of *Daahoud*[58] he spit out whole phrases of staccatissimo eighth notes.[59]

5. An astoundingly precise technique and the ability to play coherently and cleanly even at extremely fast tempos. One night club recording of *Cherokee*[60] roars along at ♩ = 410!

6. A rich, mellow tone quality that was perhaps even more beautiful than Miles Davis's, although it lacked Davis's poignancy. Bobby Shew explained to me that Brown often opened the aperture (formed by the lips) to get a big, warm, open tone on strategic notes, and then closed down again in order to play other notes more easily. The result, especially when he opened up at a phrase ending, was to create the illusion that all of his notes were equally open and full in tone quality. Mostly he played in the middle register of the trumpet, where the beauty of his tone was most effective. He could play high notes easily, however, and used them occasionally for dramatic contrast.

7. A unique vibrato that he normally used only on the last note of a phrase.

8. An unsurpassed ability to play ballads. He is famous for an entire album of ballads, accompanied by strings and a rhythm section.[61] But most of these 12 short performances are eclipsed by other ballad recordings, such as those in the list below.

9. A personal vocabulary of melodic figures. One of his favorites is a triplet-eighth phrase ending:[62]

Another, found in ballads, is a profusion of lower-neighbor grace notes; his predecessors used them as well, but not as extensively:[63]

Usually, however, Brown based his melodic ideas on the common stock of bebop figures. His genius lay more in *how* he played, than in *what* he played.

Brown made most of his best recordings with Max Roach; all but the first on this list of exceptional performances stem from that fruitful partnership:

Easy Living (ballad)—28 August 1953, Blue Note 1648
I Don't Stand a Ghost of a Chance, take 7 (ballad)—3 August 1954, EmArcy 36008
Stompin' at the Savoy—5 August 1954, EmArcy 6112
Joy Spring and *Daahoud,* 2 takes—6 August 1954; EmArcy 6075, and Mainstream 386—this version of *Joy Spring* is in F, and is much better known than the E♭ version done a month earlier
Jordu and *Parisian Thoroughfare*—30 August 1954, Gene Norman Presents 126—these concert performances are better than the more famous studio recordings done earlier that month
Star Dust (ballad)—20 January 1955, EmArcy 6101—performed by four-fifths of the Roach-Brown Quintet, plus strings
Blues Walk, take 8—24 February 1955, EmArcy 36036 (see the discussion of this great ensemble performance in Chapter 10)
Sandu (blues)—25 February 1955, EmArcy 6506
Hot House (What Is This Thing Called Love?), Woody 'n' You, and *Walkin'* (blues)—7 November 1955, Columbia 35965—the acoustical quality of these night-club performances is awful, but Brown's extended solos are excellent
Gertrude's Bounce—4 January 1956, EmArcy 36070
Pent-Up House and *Kiss and Run*—22 March 1956, Prestige 7038—this session officially was under Sonny Rollins's leadership, but it was actually the Roach-Brown Quintet
What's New (ballad)—1956, Elektra Musician 60026—recorded reasonably well on amateur equipment in a night club

On 25 June 1956 Brown played an informal concert in Philadelphia with local musicians. An audience member, using amateur equipment, recorded three pieces (*Walkin', A Night in Tunisia,* and *Donna Lee*—

issued on Columbia 32284). Brown played some excellent extended solos, and after the last piece he thanked his highly receptive audience with a short, modest speech. Then he, his pianist Richie Powell, and Powell's wife left by car to drive to Chicago for a performance. Along the way Powell's wife, who was driving, lost control of the car; the ensuing accident killed all three. And thus the jazz world lost prematurely another of its greatest musicians.[64]

The Clifford Brown legacy lives with us through his recordings, of course, and also through the extensive influence he had on other trumpeters.[65] One of Brown's brightest "students" was Lee Morgan (1938–72). Morgan first recorded shortly after graduating from high school in 1956. In those early recordings his debt to Brown was large; on *P.S. I Love You,*[66] for example, he overextended his credit by using far too many of Brown's lower-neighbor grace notes. But his talent was obvious even at such a young age, and the jazz profession noticed quickly. While still 18 he recorded three albums as leader and played on albums by Hank Mobley, Howard Rumsey, Johnny Griffin, Clifford Jordan, and Dizzy Gillespie. By age 19 he was an extremely self-assured soloist, playing Brown-inspired phrases with power and polish, but with a brasher tone and vibrato that hinted of an emerging distinct style.

In 1958, Morgan joined Art Blakey's Jazz Messengers, beginning a fruitful musical association that continued off and on until 1965. His powerful, aggressive style was the perfect complement to his leader's drumming, and some fine recordings resulted. He also made albums as a leader, two dozen for Blue Note and a few for other labels, during his 15-year recording career. *The Sidewinder,* a 24-measure, folk-like blues that received extensive playing time on radio in the 1960s, is his most famous recording. But *Gary's Nightmare,* another blues from the same album, is better, as is the stunning *Sleeping Dancer, Sleep On,* and other pieces from the album *Like Someone in Love.* The albums *Take Twelve* and *The Sixth Sense* also demonstrate his first-rank skills as a trumpeter and musician.[67]

Morgan used elements of Clifford Brown's style throughout his career; both men's styles are similar in range, tone quality, clarity of technique, lower-neighbor grace notes, and melodic phrases. But Brown had a lighter, gentler sound softened by a uniquely expressive vibrato; Morgan's style was less subtle, and he had a bigger and more open sound in the lower register. The life styles of these two trumpeters also differed markedly. Where Brown remained free of drugs, Morgan did not, and his drug use complicated his professional life. Both men met sudden deaths at an early age. Brown's was unrelated to his personal life, but Morgan was shot to death by his jilted mistress.

Donald Byrd (born in 1932) began recording in the mid-1950s. He too borrowed Clifford Brown's grace notes, melodic patterns, gentle caressing of notes, and full-tone low notes, though he also borrowed Clark Terry's lip slurs. But his tone quality in the middle and higher ranges sounded pinched and less full when compared with Brown's. A less aggressive and emotional player than Lee Morgan, many of his solos lacked drama; they stayed at one level of intensity from beginning to end. Also, his ballad playing was servicable but unimpressive.

By 1957 Byrd's tone was fuller and his playing, as in the November 15 session under Red Garland's leadership, was full of confidence.[68] Although he still owed much to Clifford Brown, eventually he moved away from his primary role model. In *I'm a Fool to Want You,* his stark, haunting sound, with its spare use of vibrato and the scarcity of lower-neighbor grace notes, has very little connection with Brown. And in *Free Form,* he dabbled briefly with Miles Davis's crying fall-offs.[69]

In the 1970s Byrd began performing heavily arranged, two-chord vamps of fusion jazz, and played mostly short, uninteresting solos over multilayered backgrounds. He profited financially with his commercial product, but lost the respect of many in the jazz world. For a time he also lost his polish as a jazz player; for example, his performances with Sonny Rollins in April 1978 (Milestone 55005) suffer from unreliable intonation and technique. In recent years Byrd has devoted less time to playing and more time to teaching.

Richard "Blue" Mitchell (1930–79) was another Clifford Brown imitator. In his early recordings, dating from the late 1950s, his full and sweet tone, with its bell-like clarity, closely resembled Brown's. As his career progressed that tone became even more bell-like, beautiful, and powerful. His other role models included Gillespie (his flawless execution of Gillespie's chromatic triplet figure and sixteenth-note figure from *One Bass Hit II*), Kenny Dorham (his zig-zagging chromatic descent), and, in the late 1970s, Freddie Hubbard (his plaintive sound—see below). Perhaps his favorite melodic trademark, which he used sparingly but always effectively, was a special treatment of the three-note chromatic descent, labeled Figure 4A on page 32:[70]

While he did not add greatly to the vocabulary of bebop, he spoke that vocabulary eloquently.

His early recordings, made when he was in his late twenties, are technically and stylistically assured and impressive. Several of the albums he made with Horace Silver have excellent samplings of his work. He was in Silver's quintet for some famous recordings, among them, *Peace, Sister Sadie, Nica's Dream, Filthy McNasty,* and *Silver's Serenade.*[71] His powerful long notes, his rhythmic drive, and his melodic inventiveness greatly enhance these and companion pieces in Silver's albums.

In the 1970s he, like Donald Byrd, took a more commercial approach to music, by recording harmonically static songs with rock rhythms and thick background textures. Unlike Byrd in recordings of this type, however, Mitchell played his solos with undiminished conviction and musicality. On *Satin Soul* and *Mississippi Jump,* for example, he "sings" his creative melodies majestically, with a beautiful, clear tone.[72] Concurrently, when working as a sideman in a variety of groups, he continued to play inspired bebop. In his last years he and Harold Land co-led an excellent quintet.

Nat Adderley (born in 1931) initially used Clark Terry and then Miles Davis as role models. His timbre is largely his own, however, in part because he plays cornet rather than trumpet. He produces a surprisingly intense sound from this inherently mellow instrument, particularly when he plays loudly in the upper register. Melodically, he leans heavily on the blues scale, almost no matter what piece he plays. He has a rich array of blues-scale melodic patterns in his repertory, and an equally varied array of expressive attacks and releases. He plays with great drive and swing, whether playing low notes softly, which he does typically at the beginning of a solo, or higher notes loudly.

Nat, although always overshadowed, contributed many hard-swinging and effective solos to the musical success of the Cannonball Adderley quintets and sextets (see Chapter 10). Nat also wrote *Work Song,* a piece that the brothers made famous. Since his brother's death in 1975, Nat has led a variety of groups.

During the 1960s, '70s, and '80s, Freddie Hubbard (born in 1938) was one of the most powerful and stylistically distinctive trumpeters and flügelhorn players in jazz. No other major trumpet soloist could match his huge, warm tone, top his flair for the dramatic, or outrun him in fast passages.

Here and there in Hubbard's recordings one may find a lower-neighbor grace note reminiscent of Clifford Brown, a mournful fall-off or a Harmon-muted phrase reminiscent of Miles Davis, or some patterns similar to those used by Blue Mitchell and others. But mostly his style is his own, and has been for decades. His most instantly recognizable melodic trademark is his extended tremolos. They span a minor third as he moves them up or down the chromatic scale:[73]

Evidently Hubbard finds the figure easy; sometimes he uses it excessively, perhaps because it may elicit cheers from a receptive audience. When he uses it judiciously, however, it adds color and drama to the climactic moments of a high-energy solo. Another Hubbard trademark, found primarily in loud and fast solos, is a series of three high notes, usually in a down-up configuration:

This figure is another dramatic gesture for a climactic solo moment. The final high note may last anywhere from a beat or two to several measures. The high pitch and long duration indicate his self-assurance as a player. Hubbard's formidable technique allows him to play scalar runs and extended running passages with great clarity and accuracy. Sometimes his elaborate melodic figuration includes multiple repetitions of a single figure. One of his favorites is a minor-seventh arpeggio:[74]

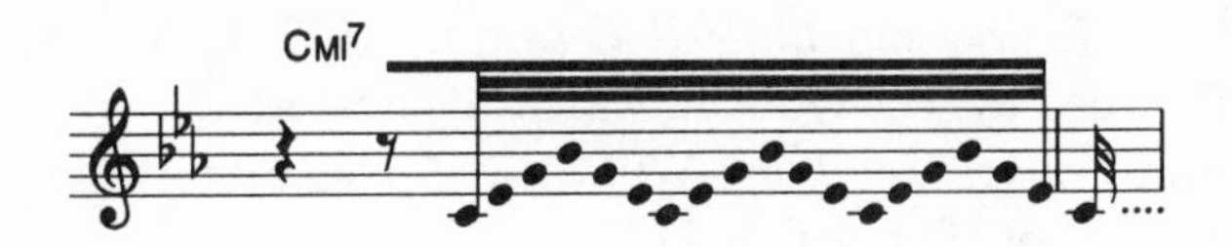

At other times he bases his repeated figure on rapidly repeated notes, using alternate fingerings, such as alternating between open position and first valve. As on the saxophone, these alternate fingerings produce slightly different notes, so the pitch fluctuates up and down slightly. Occasionally he uses a distinctive way of articulating swing eighth notes, learned perhaps from recordings by Cannonball Adderley (see page 56):[75]

(Note that this excerpt contains a figure that Kenny Dorham and others used frequently; it is an important part of Hubbard's vocabulary as well, especially in fast passages.) Finally, there is the heart-rending poignancy of moments such as the opening phrase of *Delphia* (see the list below). The subtle pitch bending and fluctuations in dynamics, elements that he manipulates masterfully in these expressive moments, defy meaningful representation in musical notation.

Since his recording debut with the Montgomery brothers in 1957, Hubbard has performed in many different musical contexts. He has played action jazz with Ornette Coleman and John Coltrane, an extended composition by Ilhan Mimaroglu, and (following the trend among many players in the 1970s) fusion.[76] But his roots are in the bebop of the 1950s, and the bulk of his extensive discography reflects those roots. On his best records he speaks the language magnificently, much as a great Shakespearian actor speaks English. His musical voice resounds deeply from such recordings as these:

Cascades and *Butch and Butch*—23 February 1961, Impulse 5—from a classic album by Oliver Nelson

You and the Night and the Music and *Interplay*—July 1962, Riverside 445—from a Bill Evans session

Pensativa—10 February 1964, Blue Note 4170—from a Blakey session

Dolphin Dance—17 March 1965, Blue Note 4195—from a session with Herbie Hancock

Delphia—January 1970, CTI 6001

Ganawa—1972, CTI 6016—from a session with Randy Weston

Third Plane, Jessica, and *Little Waltz*—July 1977, Columbia 34976—recorded by The Quintet, with Wayne Shorter, Herbie Hancock, Ron Carter, and Tony Williams, all colleagues from the 1960s

Byrdlike—2 May 1981, Pausa 7122

Bird Like (same tune as the preceding with a more appropriate spelling, since it is a blues in the Parker tradition)—29 November 1981, Fantasy 9626

Hubbard, more than any trumpeter since Clifford Brown, exerted a strong influence on younger players. Among the first-class players who have learned stylistic matters from him are Guido Basso, Oscar Brashear, Randy Brecker, Tom Harrell, Terumasa Hino, Wynton Marsalis, Blue Mitchell, Jimmy Owens, Woody Shaw, and Bobby Shew. His influence on other players varies from one to another, of course, and players often draw from more than one role model, as we have seen. Shew, for example, learned from Blue Mitchell as well as from Hubbard, and Marsalis soon turned away from Hubbard to explore other sources. Nonetheless, Hubbard, along with Gillespie, Davis, and Brown, is a principal architect of bebop trumpet playing.

CHAPTER SEVEN

Pianists

Much of the bebop idiom springs from the rhythm section—most typically a pianist, bassist, and drummer. These players are in the drivers' seats for any band. They supply most of the energy and forward momentum needed to make good jazz. And if the textures, harmonies, and rhythms emanating from the rhythm section are in the bebop style then in essence the music will be bebop, even when the soloists are playing in the swing, action jazz, or fusion idioms.

The tapestry of sound woven by the bebop rhythm section is multifaceted, for each player interlocks with his/her partners in several ways. The harmonic background usually results from a keyboardist and/or a guitarist interacting with a bassist, though sometimes one player handles the task alone. The rhythmic textures result from everyone in the section interacting with each other and with the soloist. Then there are the timbres produced by 1) the drummer shifting from cymbal to cymbal, from drum to drum, and from sticks to brushes; 2) the bassist moving from string to string and from register to register, and 3) the keyboardist and/or guitarist varying the chord voicings. Finally, each member of the section often functions as both a soloist and an accompanist.

The chord players in the rhythm sections are keyboardists—pianists primarily, but also electric organists, electric keyboardists, and synthesists—and guitarists. Of these, the largest group is the pianists. (For the other chord players, see Chapter 9; for bassists and drummers, see Chapter 8.)

Pianists deal differently with sound than do front-line players. They do not bring their instruments to the performance venue, but instead play whatever instrument awaits them. The tone of that instrument stems from many factors, most of which are beyond the pianist's control. Between the player's fingers and the vibrating strings is a long series of mechanical actions. Once the player depresses a key, s/he can do only two things: allow the sound to decay slowly over time or cut short the sound by releasing the keys. For these reasons, different players playing the same piano will produce largely the same tone quality, except for the differences they can create through different attacks and loudness levels. (One need only sample the recordings made in the 1950s by different pianists playing Rudy Van Gelder's Steinway to realize that fact.) Yet we have no trouble distinguishing Thelonious Monk from Horace Silver, Bud Powell from Oscar Peterson, Red Garland from Bill Evans. Among other things, each has a distinctive repertory of favorite figures and comping habits. Also, a good-quality grand piano is highly responsive to subtle differences of touch, so the different loudness levels of individual notes in a melody or chord can be a part of a player's style.

An important difference between solos by a swing and a bebop pianist is the rhythmic approach each takes for the left hand. To oversimplify, the swing pianist often uses his/her left hand to play a fairly continuous striding ("boom-chick") accompaniment, while the bebop pianist normally accompanies, or "comps," with occasional chordal punctuations, often between beats. The beginnings of bebop comping occur in recordings by Earl Hines, Duke Ellington, Count Basie, and other swing pianists. But younger players were the first to reject the striding left hand as a normal performance option. At first the bebop pianists continued to use swing-style chord voicings, often playing root-fifth-tenth or root-seventh-tenth voicings. By the 1950s, however, many beboppers were playing rootless voicings in the tenor register, leaving the root of the chord to the bassist and usually avoiding the bass register.

There are also melodic differences between swing and bebop pianists. Swing pianists' melodies often consist largely of chord tones played in a rhythmically varied way (♫ | 𝄾♫ ♩ ♩ 𝄽 | ♩ ♫ ♫‿♩ |), with many notes doubled in octaves or supported by chords. They often embellish phrases with rapid arpeggios, or pentatonic runs, spanning several octaves. Bebop pianists' melodies, like those by their horn-playing colleagues, are often more scalar than chordal, and normally include long strings of eighth notes, two and three per beat, that incorporate the melodic formulas and unpredictable accent patterns of bebop. Octave and chordal reinforcement of melody tones is less common, as are extended arpeggios and scalar runs, than they are in swing. These changes in melody and accompaniment result in a strikingly different piano idiom. And since comping, not stride style, is the norm during theme choruses and other players' solos as well as during piano solos, the bebop pianists have rewritten the job description for the entire performance.

Thelonious Sphere Monk (1917–82) was a boundary stretcher if there ever was one. No other major bebop pianist had such an unorthodox style or used such unusual piano techniques. No other bebop composer of the 1940s wrote such simplistic melodies on the one hand and such rhythmically and intervalically complex melodies on the other. No other bebop musician behaved so enigmatically, both on stage and off. Since the subject of this chapter is pianists, I will focus just on his playing here, but everything about his music and his personality was of a piece. He was at his best playing his own compositions with his great quartets of the late 1950s and 1960s; he was the single most important composer in bebop; and his quirky, complex music was a perfect reflection of his personality.[1]

Monk was a bebop pioneer. He was the house pianist at Minton's Playhouse in the early 1940s, and nightly collaborated with the other young players who were unconsciously developing the new style. In 1944 he worked with Coleman Hawkins, with whom he made his first studio recordings, and he played briefly in Gillespie's big band of 1946. His unorthodox playing style puzzled or disturbed most players and listeners at first, however, so he performed in public only sporadically during the 1940s and early 1950s.

By the middle 1950s his fortunes began to change. His great ballad *'Round Midnight* (first recorded by the Cootie Williams big band in 1944) and his fast *Fifty-second Street Theme* had become jazz standards, and players were performing some of his other pieces occasionally. His public performances and recordings gradually increased in number. By the late 1950s the jazz world viewed him as a major player and composer. *Time* magazine even put his picture on the cover of the 28 February 1964 issue. His quartets were among the world's premier bebop groups; during the years 1955–70 his sidemen included saxophonists Sonny Rollins, John Coltrane, Johnny Griffin, and (during 1959–70) Charlie Rouse, bassists Wilbur Ware, Ahmed Abdul Malik, Sam Jones, John Ore, and Larry Gales, and drummers Art Taylor, Art Blakey, Shadow Wilson, Roy Haynes, Frank Dunlop, and Ben Riley. These groups swung with such power and grace that Monk often left the piano to do an eccentric dance while his inspired sidemen played. By the early 1970s Monk's health began to deteriorate. In 1976 he stopped playing in public, and spent his last years in voluntary isolation.

Monk's piano style at first seemed to come from nowhere. It had an awkwardness and angularity that had little obvious connection to either his swing predecessors or his bebop contemporaries. But though he did rework piano technique, melodies, chord voicings, and comping rhythms, he retained at least three elements from earlier players: his frequent use (compared to other bebop pianists) of the striding left-hand accompaniments, some filigree runs of Tatum and Wilson, and the powerful and dissonant comping occasionally used by Ellington.

Monk's usual piano touch was harsh and percussive, even in ballads. He often attacked the keyboard anew for each note, rather than striving for any semblance of legato. Often seemingly unintentional seconds embellish his melodic lines, giving the effect of someone playing while wearing work gloves. These features were a result of Monk doing everything "wrong" in the sense of traditional piano technique. He hit the keys with fingers held flat rather than in a natural curve, and held his free fingers high above the keys. Because his right elbow fanned outward away from his body, he often hit the keys at an angle rather than in parallel. Sometimes he hit a single key with more than one finger, and divided single-line melodies between the two hands.

The preceding discussion might suggest that Monk was an untutored amateur at the keyboard. But in an instant he would switch from this unorthodox technique and execute a run or arpeggio with dazzling speed and unerring accuracy. In ballads he often decorated long notes with an octave tremolo played cleanly and extremely rapidly. Martin Williams (1970: 118) and Ran Blake (1982: 27–28) have pointed out that he had a well-developed finger independence, so that he could play a trill and a melodic line simultaneously with his right hand. And although his normal touch was a hard-edged attack, he occasionally played in a gentle, legato manner. Clearly he had a considerable command of piano technique.

While Monk's one-of-a-kind attack is itself a clear identifier of his style—it colors nearly everything he played—he had some favorite melodic devices, too. Some characteristic melodic patterns derive from the whole-tone scale. Descending and ascending multi-octave whole-tone runs occur commonly, especially in his earlier recordings. This whole-tone segment is a simpler figure that sits under the hand:[2]

This whole-tone figure is a series of augmented triads (the only possible triads in a whole-tone scale), played in a 3-against-2 or 3-against-4 rhythm:

Two of his signature figures exploit half-steps. One is a measured trill with a rising ending:

(This figure became a regular feature of his *Little Rootie Tootie* solos.) The other is one that most players would play in this straightforward way—

—but Monk played it this way:

Another of his figures fits the hand easily; after using it in improvisations throughout his career, he made it the insistently repeated motive of his blues theme *Raise Four,* the title of which refers to the augmented fourth, E:

In some solos the intervals shown here—a perfect fourth and an augmented fourth—become two perfect fourths, or a perfect fourth and a perfect fifth.

Monk enjoyed the sound of parallel sixths. He built them into the themes of *Misterioso, Crepuscule with Nellie,* and *Functional,* and he often improvised long passages—up to a chorus or more—in parallel sixths. He also unified long sections of his solos with a single simple motive. In the master take of *Bags' Groove*[3] he builds his first chorus on this sparse motive:

Later, after developing a few equally simple motives, he returns to this rhythm for his penultimate chorus.[4]

We identify Monk almost as much by what he did not play as by what he did play. Many of his solos contain numerous rests and long notes. (The most extreme example occurs in his quirky solo on *The Man I Love,* take 2,[5] where the gaps are so long that Miles Davis felt compelled to jump in and play to keep the music moving.) And because he liked melodic space he seldom played long eighth-note strings. Consequently, he played few bebop clichés; Figure 1A (see p. 31) is the only Parker figure that he used extensively.

Monk avoided quoting popular song themes almost entirely. Often, however, he quoted himself, usually by incorporating the theme of the moment in his solo. And sometimes one of his other pieces pops up in a solo; for example, a quotation from *Misterioso* appears in some of his *Straight, No Chaser* solos (both are blues in F). During his solos he comped less than Bud Powell, Al Haig, and others, largely because he was using both hands to play his melodies. When he did play an accompaniment he often reverted to simple chords played in the stride style of his youth.[6] In some of his solo recordings the striding accompaniment is constant, chorus after chorus. He was unconcerned that this texture was largely out of favor among his bebop colleagues. On the contrary, he once commented proudly to his record producer that he (Monk) sounded "just like James P. Johnson" (Keepnews 1988: 124). Often he would stop comping altogether for his sidemen, relinquishing that job to his bassist and drummer, either to create some textural contrast or so that he could move around the stage.

When performing with his quartets, Monk usually put himself in a subordinate position. The group was the Thelonious Monk Quartet but it *sounded* like the Johnny Griffin Quartet or the Charlie Rouse quartet, because the saxophonist usually soloed first and longest. Monk's solos were generally shorter and less dramatic. Perhaps he viewed himself primarily as the group organizer and composer, not the star player. Surely his compositions were important to him, for often his short solos would be theme choruses or theme paraphrase choruses instead of independent improvisations, and even in solos in which he strayed from the theme he normally returned to it during the final measures.

Widespread recognition came late for Monk. All but totally ignored as a role model in the 1940s, he was copied in the 1950s and 1960s, but mostly by players wishing to learn his chord voicings, not his improvising style. Horace Silver copied Monk's octave tremolo and some of his keyboard technique, but little else (see below). To sound like Monk one must approach the keyboard as Monk did, and few have been willing to abandon their more orthodox playing techniques. Marcus Roberts can come close to Monk's sound and improvisation style, but even he is far from being a slavish imitator. Monk, more than any other major figure in bebop, was, and remains, an original.

Lennie Tristano (1919–78) was another one-of-a-kind pianist in 1940s bebop. Both Monk and Tristano were composers who recorded occasionally, and usually as leaders. Both tended to comp as though oblivious to the soloist.[7] There the similarities end, however. Tristano had a more traditional European technique and a lovely, singing tone, and favored long, legato melodic lines in his solos. His compositions, radically different from Monk's, were melodically and harmonically among the most complex of the early bebop tradition. And unlike Monk, he devoted much time and energy to private teaching.

From the mid-1940s, Tristano was the central figure in a small bebop subgroup that included alto saxophonist Lee Konitz, tenor saxophonist Warne Marsh, guitarist Billy Bauer, and pianist Sal Mosca. They favored unusually angular and unvocal themes, improvised melodies filled with almost arbitrary chromaticisms that often obscured the basic harmonic structure, legato phrasing nearly devoid of sharp attacks and accents, and simple time-keeping with brushes by their inconspicuous drummers. The jazz world regarded them as "cerebral" and "cool." The Tristano Sextet's unstructured and undramatic collective improvisations, *Intuition* and *Digression* (see pp. 22–23), and Tristano's overdubbed *Descent into the Maelstrom* are surprisingly early examples of action ("free") jazz.[8]

Although his most famous recordings date from the 1940s, Tristano's work a few years later is superior. His style lost its radical aspects and he adapted more to the bebop mainstream, but he retained his attractive habit of spinning motives out over several measures. *Line Up* (an *All of Me* contrafact) and *East Thirty-second,* both from 1955, are wonderful one-hand improvisations in the bass and tenor registers of the keyboard. In both pieces his articulation and phrasing are much more varied and interesting than they were in his early recordings. *All the*

Things You Are, also from the mid-1950s, is a lengthy performance full of invention. And *C Minor Complex* is an impressive solo performance in which he plays walking bass lines to accompany his inventive melodies.[9]

The personal tragedies suffered by some musicians is an unfortunate leitmotif of this book. Tuberculosis, drug and alcohol addiction, cancer, automobile accidents, and even murders have cut life short for many of bebop's greatest creators. This tragic leitmotif applies to early bebop's most influential pianist, Earl "Bud" Powell (1924–66). In his case, mental illness coupled with alcohol and other drug abuses resulted in a life punctuated by lengthy stays in mental hospitals and filled with unreliable professional and personal actions. Given the difficulties he faced it is remarkable that he could function as a musician. Sometimes, in fact, he could not play at all; other times he played, but in a tortured, struggling manner that was painful for the listener and probably shattering for him. When he was temporarily in control of those inner demons, however, he could play magnificently:

> One night in Paris, in '58 or '59 . . . Ray Brown and Herb Ellis and I went down to the Blue Note. Bud was playing there with a French rhythm section. . . . He got up at the piano, and proceeded to play like you never heard anyone play before. No one ever heard better choruses than we heard that night. Ray Brown said, "Man, I worked with him on Fifty-second Street, I worked with him with Bird, he never played better than this." He was playing *The Best Thing for You* at a tempo you wouldn't believe. That thing is a roller-coaster of [chord] changes . . . and he did it like a loop-the-loop, chorus after chorus, relentless, with such strength. . . . He never ran out of gas. It was a night that I'll never forget! . . . [That] night . . . was the greatest jazz performance I have ever been lucky enough to hear. . . . (Lou Levy, quoted in Lees 1990: 5)

A move to Europe in 1959 gave him short-term peace and stability, but mental and physical problems plagued him there as well. He returned to New York City in 1964 and declined steadily until his death two years later.

There was an important symbiotic relationship between Monk and Bud Powell. In the early 1940s Monk often brought Powell into Minton's and gave him the opportunity to play and be heard. He also spent much time with the teen-aged Powell, teaching him informally. He honored Powell with his composition *In Walked Bud,* which became a jazz standard, and wrote other pieces with Powell in mind as the performer. Powell, in turn, when he was in Cootie Williams's band, persuaded Williams to premiere *'Round Midnight* on record. Later, Powell was one of the first pianists other than Monk to play Monk's music on his own sessions. In 1947 he recorded Monk's *Off Minor* nine months before Monk recorded it.

Powell was one of "Bird's children" as surely as were any saxophonists of the time. He, more than any other early bebop pianist, transferred Parker's melodic vocabulary and phrasing to the piano. He used nearly all the Parker figures that I show in Chapter 3, and several that I did not show. He also based sections of his solos on descending scales, just as Parker did. Powell began to imitate Parker early; his 1944 recordings with Cootie Williams contain several Parker figures. Since these recordings predate Parker's first mature-style studio recordings, it might appear that the two men arrived at their styles independently. But Powell had played in jam sessions with Parker before making these recordings, and probably learned some of Parker's ideas through first-hand contact.

Powell also borrowed heavily from Art Tatum. Several of Tatum's ornate runs were in his vocabulary, especially during the 1940s. And while Powell normally used a left-hand style that was far different from the swing-style norm, he was adept at playing a striding left hand, and did so occasionally throughout his career.

His best solos were impressive, especially during the 1940s, when no other pianist had his command of the bebop idiom. He played his long, fluid melodic lines in a powerful, aggressive manner, using a variety of accented and unaccented notes much in the Parker manner. But in imitating Parker and Tatum, two of the foremost virtuosos in jazz during the 1940s, Powell set his sights high and made heavy demands upon himself. He seemed to push himself to the extreme limits of his technical ability, and sometimes beyond it, which led to sloppily executed sixteenth-note runs, runs suddenly aborted, obvious mistakes in theme statements, flubs in improvisations caused by inaccurate fingering, and an uncontrollable, non-stop grunting with which he accompanied many solos.[10] Listening to some fast pieces, such as the May 1949 recording of *All God's Children Got Rhythm,* is an anxiety-inducing experience. Can he stay with the bassist and drummer? Will he play himself into a corner and have to stumble on somehow? Perhaps his most ill-advised recording was his amateurish rendering of C.P.E. Bach's *Solfeggietto,* renamed *Bud on Bach.* (The subsequent improvisation is better.)[11]

These technical shortcomings aside, Powell was a major figure in early bebop, whose best recordings were inspirational to many other pianists. According to Kenny Clarke, Thelonious Monk wrote his early pieces for Powell, "because he figured Bud was the only one who could play [them]" (Gitler 1985: 102). Among Powell's best recordings are the following:

Buzzy, take 5 (blues)—May 1947, Savoy 652—with Parker
'Round Midnight and *Ornithology (How High the Moon)*—30 June 1950; Le jazz cool 101—from a Birdland broadcast with Parker
The Fruit—February 1951, Clef 610—from a solo piano session
Un poco loco, take 3—1 May 1951, Blue Note 1503 (see Chapter 10)
Woody 'n' You—21 March 1953, Session 109—from a Birdland broadcast
Lullaby of Birdland, alternate take, and *Woody 'n' You*—5 April 1953, Elektra Musician 60030—from a night-club performance in Washington, D.C.
Hot House (What Is This Thing Called Love?)—15 May 1953, Debut 4—from a concert performance with Parker at Massey Hall, Toronto
Good Bait—15 December 1961, Columbia 35755—with Don Byas in Paris
Off Minor and *No Name Blues*—17 December 1961, Columbia 36805
Rifftide (Oh, Lady Be Good), Straight, No Chaser (blues), and *Hot House*—26 April 1962, Delmark 406
Blues for Bouffemont—31 July 1964, Fontana 901

Powell's melodies, of course, attract our attention more than his accompaniments. But he and other bebop pianists established new ways to use their left hands. Powell's solo recordings of 1951 offer a good chance to hear his accompaniments, for he played in the same manner in these pieces as he did with a bassist and drummer. The example shown below comes from *The Fruit,* one of his lesser-known compositions (this excerpt is from the bridge of the second chorus). It is a fine piece and his performance of it is fascinating.

Nearly every sound his left hand makes includes a chord root. Typically he plays the root and the tenth or root and seventh, as he does throughout this example. Other voicings, not shown here, include root-fifth-tenth, root-fifth-seventh, root-third-seventh, and root fifth. Sometimes he plays only the root, always in the string bass range. Most of these left-hand sonorities are syncopated, and usually anticipate beat 1 or 3. Overall, he articulates the beat in his right hand more than his left.

His phrasing is remarkably similar to Parker's. Except for the sustained D♭ in the second full measure, the entire passage—with its weak-to-strong phrasing of swing eighth notes, off-beat accents, melodic shape, and decorated scalar descent—could have occurred in a Parker solo. The combination of this Parker-style melody and his left-hand punctuations is the distilled essence of the early bebop ensemble idiom.

Nearly every important bebop pianist in the 1940s and early 1950s played with Charlie Parker. Some, such as Erroll Garner, Lennie Tristano, and Oscar Peterson, played on only a few record dates, broadcasts, or concerts; others, such as Bud Powell, had much longer professional associations. Al Haig (1924–82) was in this latter group. He began playing with Parker and Gillespie in 1945; after playing on the famous *Salt Peanuts* and *Hot House* recordings (see pp. 15-17), he traveled with them to Los Angeles. He returned to New York City in 1946, while Parker remained on the West Coast, but rejoined him during 1948–50.

Haig was a superb accompanist, and was much better at it than Bud Powell, whose comping often sounded mechanical and too legato to generate much forward momentum. He was also a capable soloist. His technique was superior to Powell's and from the late 1940s on he handled any tempo and harmonic context with ease. Stylistically he owed much to Powell, but was no slavish imitator.

While Haig may have been the better pianist, he was not necessarily the better musician. In Powell's solos we can hear the sweat and strain and can sense the intense struggle and commitment to the music. Haig, on the other hand, seemed to breeze effortlessly through his solos; he needed no toweling off afterward, for he had put little of himself on the line. Powell's solos are technically flawed but emotionally strong statements; Haig's are technically polished but facile, unemotional statements.

In his later recordings Haig's emotional breadth expanded, especially in some lovely ad-lib tempo ballads. Also, he abandoned the Powell-derived left-hand style of playing roots in the bass range. Instead, he moved his left hand into the middle range for close-position chords, thereby clarifying the harmonies and giving his bassists more maneuvering room.

Most early bebop pianists established themselves in the swing idiom initially. The oldest, Mary Lou Williams (1910–81), built her reputation as Andy Kirk's pianist from 1929 to 1942, and as a top big-band arranger, writing for Kirk, Benny Goodman, Earl Hines, Ellington, and others. Years of close contact with younger players on 52nd Street in New York City during the 1940s inspired her to modify her style. From then on her playing, arranging, and composing were mostly within the bebop idiom. Throughout her career she adapted easily to newer musical trends, adding fourth chords, ostinato compositions, and asymmetric meters to her vocabulary. Her stylistic evolution was continuous and dramatic.

Jimmy Rowles (born in 1918) had a fluid, Teddy Wilson-inspired swing style during his years with Lester Young, Benny Goodman, and Woody Herman in the 1940s. Since the 1950s his style has been essentially bebop, though he employs a variety of left-hand devices, some of which recall his early years in swing. Rowles's encyclopedic knowledge of popular songs (including the verses), his exquisite touch, and his unerring support of soloists make him a great accompanist. His work with Billy Holiday, Ella Fitzgerald, Carmen McRae, Sarah Vaughan, and other singers is exemplary; so is his instrumental ensemble work, such as the great album he made with Ben Webster in 1960 (Contemporary 7646). His droll sense of humor surfaces in theme quotes, in humorous and sometimes lengthy motivic developments, and in some compositions, such as his country-and-western classic, *The Ballad of Thelonious Monk.*[12]

Finding the proper stylistic pigeonhole for Erroll Garner (1921–77) is challenging. Some of his melodic formulas and much of his harmonic vocabulary was congruent with bebop, but his rhythmic vocabulary, and especially his left-hand style, was closer to swing. His early recordings with Georgie Auld and Charlie Parker show that he could have become a conventional bebopper.[13] Instead, he found a different path, and by the late 1940s developed an instantly recognizable piano style. He often played melodies in octave-spanning chords and in separately articulated eighth notes, played far behind the beat and supported by heavy-handed four-beat chording. He needed no rhythm section, but usually used one anyway, to play quiet, unobtrusive (and for the players, probably boring) accompaniments.[14] In spite of the fullness of his chords and the nearly constant rhythmic tension between his hands, his style was lighthearted and effervescent. Listeners often find themselves involuntarily smiling or even laughing with pleasure while listening to his rhythmic tightrope walking.

Billy Taylor (born in 1921) moved to New York City in the early 1940s, and immediately plunged into the musical activity on 52nd Street and the jam sessions at Minton's. His early recordings show him mixing Tatum-like runs and accompaniments with bebop melodic figures. His comping behind the soloists placed him firmly in the bebop camp, however. In fact, he developed a distinctive comping style, using block chords in both hands:[15]

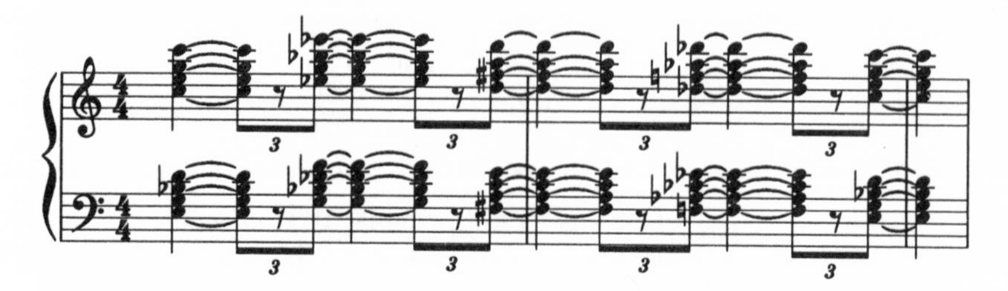

During the 1940s and early 1950s Taylor was a sideman for many different swing and bebop leaders, and led groups of his own. His favorite ensemble size has been the trio; among his best have been those with Ben Tucker and Grady Tate (1967, Tower 5111) and especially with Victor Gaskin and Bobby Thomas (June 1988, Taylor-Made CD 1001). Although his mature style is clearly bebop-based, his activities as a world-renowned jazz educator led him to master earlier styles as well.

Londoner George Shearing (born in 1919) was a swing-style player until the late 1940s when he moved to New York City. Then he quickly adopted much of the bebop vocabulary, as both a soloist and a comper, always playing with a reserved, gentle touch. His facile technique sometimes outran his ability to imagine meaningful melodies, but the same technique allowed him to play locked-hands style—first explored by Milt Buckner, Dodo Marmarosa, and others—almost as fast as others played single-note lines.

In 1949, Shearing formed a quintet with a suitably soft instrumentation: vibraphone, amplified guitar, piano, bass, and drums (usually restricted to brushes). His quintet arrangements included locked-hands voicings with the vibraphonist doubling the top note and the guitar doubling the bottom. This "Shearing sound" was extraordinarily successful; it is the sound by which many fans identify him even today, although he abandoned the quintet format in the 1970s. The group's subdued "cocktail bar bebop" often left little room for improvisation, especially for his vibraphonists and guitarists, and his drummers probably left their sticks at home. Yet the group could play lively music, especially when joined by some Latin percussionists. In 1952, the quintet recorded Shearing's best-known composition, *Lullaby of Birdland,* which quickly became a jazz standard.[16]

Since the late 1970s Shearing most commonly has performed with just a bassist. His mature melodic style is bebop-based, but his accompaniments are more active than that of most bebop pianists, and his comping style is closer to swing than to bebop. Both as a soloist and an accompanist, he is at his best playing ballads, which showcase his beautiful, lyrical touch.

Hank Jones (born in 1918), the oldest brother to brass player Thad Jones and drummer Elvin Jones, was another who began as a swing-style pianist. But soon his role model became Bud Powell, as he adopted a similar melodic style and bass-register left-hand style (though he sometimes moved his chords into the middle register). His technical facility was greater than Powell's, but less facile than Haig's. In the 1940s he played with Hot Lips Page, toured with Jazz at the Philharmonic groups, and began a five-year tenure as Ella Fitzgerald's accompanist. Since the 1950s he has been a freelance player; for years he was the company pianist for Savoy records and played in a much-recorded group known unofficially as the "New York Rhythm Section"—with guitarist Barry Galbraith, bassist Milt Hinton, and drummer Osie Johnson (Hinton 1988: 243). His comping skill—illustrated in Chapter 2, page 26—manifested itself by 1947, in J. J. Johnson's *Down Vernon's Alley.*[17] His full, supportive accompaniments often include brief passages of thick, richly-voiced substitute harmonies. In the 1950s he lightened his accompaniments during his solos, playing so sparsely and lightly that essentially he was a monophonic player much of the time.

Over the years Jones's musicality has grown dramatically. In the 1940s he was a good piano player with an immature style; in the 1950s he was a solid bebopper, but others with more distinctive personal idioms or flashier technique overshadowed him. His mature style, especially since the 1970s, is a treasure. Long a consummate accompanist, he became a consistently inventive and swinging soloist. His melodic lines are seldom flashy but have a compelling logic and an effortless flow. With his touch, his time sense, his humorous theme quotes, and especially his consistently high level of invention, he represents the essence of great bebop.

The same laudatory remarks apply to Tommy Flanagan (born in 1930); in fact, Jones and Flanagan often come to mind in the same thought, since both spent many early years functioning primarily as sidemen, both are superb accompanists, and both have earned worldwide recognition as superior bebop pianists. Flanagan, twelve years

Jones's junior, began his recording career in the mid-1950s, working with the cream of bebop players on dozens of albums. He spent most of the 1960s and 1970s working as Ella Fitzgerald's accompanist. Since 1978 he has worked primarily as a trio leader. His recent groups—with bassist George Mraz and drummers Kenny Washington or Louis Nash—are among the world's finest.

When Hampton Hawes (1928–77) began recording in the late 1940s, he was a heavy-handed, unpolished player, but he improved his style dramatically in the early 1950s. His eclectic, polished style included many figures used by Parker and Powell (but he played with a cleaner articulation than Powell), some Oscar Peterson phrases, and later, some Bill Evans phrases (see below), and an impressive locked-hands style in which the top notes always sang out clearly. He was among the first bebop pianists to develop the double-note blues figures and rhythmically compelling comping style that Horace Silver and others were to use in the mid-1950s. Although he was a self-taught pianist, he had great facility with rapid runs and a versatile control of touch. He played the piano (and, in the 1970s, the Fender-Rhodes keyboard) with an impeccable drive and swing from the mid-1950s until his fatal stroke at age 48. Unfortunately, his was another drug-plagued life; prison sentences for drug-related offenses interrupted his promising career several times.[18]

William "Red" Garland (1923–84) was old enough to have been among the first wave of bebop pianists. But though he played professionally in the late 1940s, his recorded legacy (apart from an obscure session with Eddie "Lockjaw" Davis) begins only in 1955, when he began his three-year engagement with Miles Davis. During the years 1955–62 he played on dozens of albums, nearly all for Prestige, and nearly always with bassist Paul Chambers and drummers Philly Joe Jones or Art Taylor. In the mid-1960s he left New York City and returned permanently to his native state, Texas.

Garland was partly responsible for the "Miles Davis sound" of the late 1950s (see page 117). He was a great team player. His gently prodding style of comping, perhaps derived from Ahmad Jamal's during the same period, complemented perfectly the muted, understated lines Davis played. In his solos he continued his leader's subtle swing, playing precisely and in a relaxed and quiet manner. When playing single-note melodies he often comped on the last part of the second and fourth beats—| 𝄽 ³𝄾 𝄾 ♪ 𝄽 ³𝄾 𝄾 ♪ |—as though gently pushing his melody forward. His best-known effect was a rich block-chord texture, again played in a calm, relaxed manner:[19]

In the 1950s writers labeled several pianists as "funky" or "soul" pianists, because they often used the techniques of gospel and traditional blues pianists. Beboppers all, they were simply augmenting their bebop vocabularies with additional expressive devices when the piece at hand contained both folk and bebop features. Gene Harris, Ramsey Lewis, Les McCann, Horace Silver, and Bobby Timmons were among the most prominent folk-influenced pianists, although most bebop pianists dipped into the folk tradition whenever it suited them.

Horace Silver (born in 1928), is famous for his simple boogie-woogie figures and straightforward, symmetrical solo melodies.[20] His first recordings, from the early 1950s, show the influence of Bud Powell, especially in pieces having complicated themes. But by the middle 1950s Powell's influence had faded. Monk also was an influence, most obviously in Silver's fast tremolo and dry, almost pedal-free ballads, but also in the stiff-fingered, percussive technique that both men share—the oddest approach to keyboard attack and fingering in jazz.

If there are superficial similarities between Silver's style and Monk's there are far more differences, in both tone quality and the notes and rhythms each used. There is a fascinating recording of *Misterioso* dating from a Sonny Rollins session (14 April 1957; Blue Note 1558), in which both pianists play. Monk comps then solos during the first few minutes; then Silver takes over, also comping and soloing. The change of style is immediate and obvious: touch, rhythm, harmony, and especially melody are all different.

For the most part Silver avoids long melodic strings of eighth notes. Instead, he favors shorter phrases in which eighth notes and longer note values commingle. These short melodies are tuneful, unfolding in easily assimilated, almost predictable ways. The motive and phrase structure in his solos often seems as carefully crafted as a song. (In fact, often it *is* a song, for he and Dexter Gordon may be the champion theme quoters in jazz. He even quoted *Oh You Beautiful Doll* in his *Quicksilver* melody.) Perhaps this compositional feature of his solos derives from his long and tireless activities as a composer. He is one of the two or three principal composers in bebop, an aspect of his music discussed in Chapter 10.

Silver plays a direct and engaging solo in *Sister Sadie.*[21] In the excerpt below are several of his solo traits: the short, simple phrases that all derive from the three-beat figure ♩ ♩ | ♩, or a variant of it; the pianist's "blue fifth" (those rapid slurs up to D); and the low tone cluster used strictly as a rhythmic punctuation:

Ever since the late 1950s Silver has used the low cluster effect frequently, especially in fast blues. Perhaps it evolved from his youthful habit of playing a chordal root or fifth plus its lower neighbor tone in the bass register. Later, when he moved his left hand into the middle register, he began to jump down quickly for a single bass-range punctuation, and those punctuations gradually changed from one or two notes into clusters.

The minor-pentatonic (1-♭3-4-5-♭7-8) and blues (1-♭3-4-#4-5-♭7-8 or 8-♭7-5-♭5-4-♭3-1) scales are among his favorites; many solos contain passages such as the following:[22]

In this excerpt, his handling of scale degrees #4 and 5 (C# and D) is perhaps a carryover from Monk's method of playing minor-second figures. Again, there is that low cluster punctuation.

While the folk-like qualities of his music are the best-known, the full scope of his melodic and harmonic vocabulary is much broader. Even when playing a folk-style blues based upon the most basic harmonies he may sequence a figure up or down through the chromatic scale, or create a bitonal effect by emphasizing the ninth, raised eleventh, and thirteenth of an enriched dominant seventh chord (i.e., emphasizing C, E, and G when the basic chord is B♭7). Many of his compositions contain no folk blues or gospel music elements, but instead have highly chromatic melodies supported by richly dissonant harmonies (i.e., *Nica's Dream, Ecaroh, Barbara,* and others). In his solos on these pieces he generally avoids using the blues-derived devices he favors in the simpler pieces.

Silver's comping for trumpeters and saxophonists is a distinctive aspect of his playing. Rather than reacting to the soloist's melody and waiting for melodic holes to fill, he typically plays background patterns similar to the background riffs that saxes or brasses play behind soloists in big bands. This trait may be the composer in him surfacing. Whatever the motivation, he works harder to provide his soloists with a stimulating, swinging background than many pianists do during their own solos.[23]

While some of Silver's fingerings would throw Miss Tillie the piano teacher into a tizzy—

—they help him create an intensely personal piano style. As with Monk, his technique may seem primitive, but it is not. He can play fast, intricate runs when he wishes. Like Sonny Rollins, Silver has found a way to make his instrument *sound* loud without actually playing loudly, for his touch is surprisingly gentle and controlled. In a real sense he is a virtuoso, albeit of a completely different sort than Oscar Peterson (see below). Because of his unique virtuosity, no one can play the blues (such as *Filthy McNasty* and *Señor Blues*) with the down-home sound more convincingly or effectively.[24] No one.

Oscar Peterson (born in 1925) is surely the world's most famous living jazz pianist. Jazz impresario Norman Granz helped him immeasurably in achieving his fame; Peterson was a local musician in Montréal until Granz added him to his Jazz at the Philharmonic roster in 1950. The wide exposure that the JATP tours gave him, and a flood of recordings on Granz's labels (Clef, Norgran, Verve, Pablo), helped even more. All the exposure in the world, however, would have helped little if Peterson had lacked the talents and abilities needed for Granz's world series of jazz. Peterson's musical skills are abundant. There is no more powerfully swinging pianist, and probably no one with a greater technical command of the instrument. Many consider him the Art Tatum of bebop. The parallel is not without merit; both men possess great technical expertise and have an easy control of amazingly fast playing, and both men are masterful solo pianists. Also, both men triumphed over physical handicaps; Tatum was legally blind, and Peterson has suffered from an inherited arthritis of the hands (!) throughout his professional life (Lees 1988: 7).

But there are more differences than similarities. Tatum was a swing player, whereas Peterson, who began in the early 1940s in the swing style, switches easily back and forth between swing and bebop. Tatum rarely strayed far from the melody of the piece he was playing; his most ornate embellishments usually appear at the phrase endings, where the themes come to rest on long notes. Peterson normally departs from the melody immediately after the theme chorus, and is likely to play his dazzlingly fast lines anywhere in the song structure. Tatum preferred to play alone, and did so for most of his career. Peterson is equally at ease in group situations and as a solo pianist. Tatum was too assertive to be a good accompanist; he seemed to compete with the soloist he allegedly was accompanying. Peterson, especially during the 1950s and 1960s, has been a fine accompanist for both singers and instrumentalists.

Peterson actually learned more from Nat Cole than from Tatum, especially in his early years. He drew upon Cole's touch, phrasing, melodic ideas, and approach to locked-hands playing. During 1951–58 he used the Nat Cole Trio instrumentation of piano, guitar, and bass. And his singing style is a blatant imitation of Cole's.

Early in his career Peterson's musical imagination was stimulated by the new style of the 1940s. By the 1950s he had perfected a stylistic blend; his boogie-woogie and striding left-hand patterns and his right-hand riffs and arpeggio-based figures hearkened back to swing, while his rhythmically compelling and powerful comping and his use of Parker-like melodic figures created high-level bebop. In the 1950s and 1960s he used boogie-woogie and stride patterns sparingly, but since the 1970s he has used them more frequently. Many times these devices occur in solo and duet recordings, where the absence of a rhythm section probably inspires his use of the fuller swing-style accompaniment textures.

Peterson's introductions and codas often contain dazzling two-handed runs played two octaves apart. The main parts of his solos abound with more florid embellishments, such as this figure, which usually appears in pieces faster than ♩ = 200:

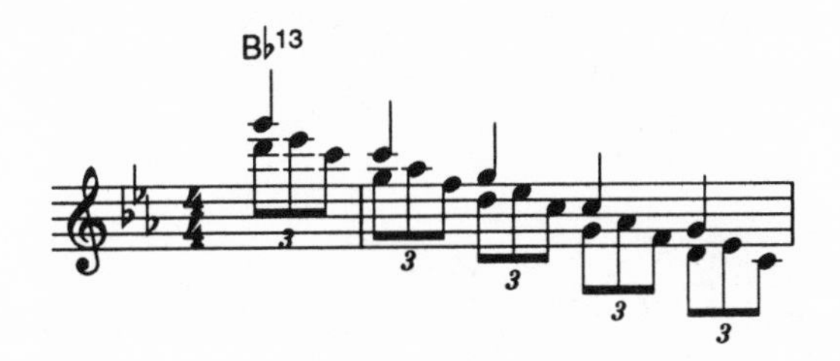

On the other hand, he also plays lines as sparse as those by Basie (from whom he borrowed a few licks). And he often fills his solos with the most basic piano blues figures. But he brings an extraordinary fluidity and control to these phrases, for his right-hand finger independence permits him to emphasize unfailingly the lower melody in the two-part counterpoint of these blues licks:[25]

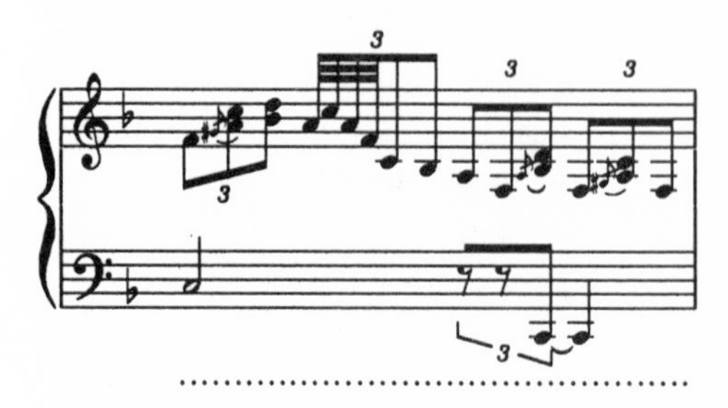

Few jazz pianists regularly use as wide a range of dynamics and touches as Peterson. In climactic moments during concerts he plays so powerfully that by intermission the piano's unison strings are hopelessly out of tune. Yet he can be the gentlest of pianists, caressing the keys to produce an exquisite tone, as in the stunningly beautiful *Little Girl Blue*.[26]

His locked-hands passages are often as fast as the single-note lines of other pianists. In such passages he may play the melody in both his left hand and the upper fingers of his right hand, which is the common practice. Or he may put the left hand a tenth below the melody note, giving rise to a richer harmonic texture, as in this $V^{7(\flat 5)}$ elaboration, derived from the diminished scale C-D♭-E♭-E-F#-G-A-B♭-C:[27]

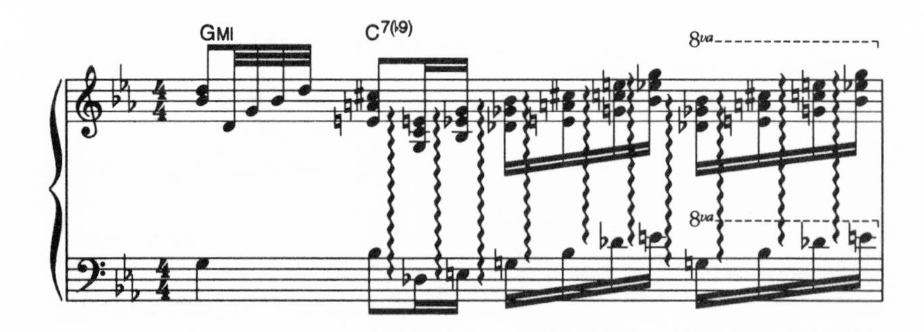

In the mid-1950s, the quiet, self-effacing Bill Evans (1929–80) began his recording career. By 1960 he clearly was a major figure in the jazz world, and long before his death he had garnered a prominent place among the most admired and imitated pianists in jazz history. His touch, as appropriate for a Chopin *Ballade* as for a Rodgers and Hart ballad, was among the most lyrical ever heard in jazz. In his left hand that refined touch turned his pungent chord voicings into elegant cushions for his right-hand melodies. His pedaling, which he used to produce an unblurred legato line, was unmatched. He never thundered, even in the most aggressive of groups and pieces. Yet with his control and ready use of crescendos and decrescendos he could be intensely dramatic, even in ballads. And he could swing magnificently; he proved that repeatedly during his tenure with Miles Davis during 1958–59, when he blended effortlessly with the most powerful rhythm players in jazz. He proved it in a different way later, when his bassists and drummers would avoid straight-ahead time-keeping and play instead melodic lines and splashes of color, leaving Evans to create the swing feel. He was a superb ensemble player; his sensitivity to the harmonic and rhythmic needs of Miles Davis and other soloists was so keen that "comping" seems a jarringly crass and inappropriate word for his impeccable accompaniments. His solo style, the focus here, was continually fresh and exciting in the early years of his career; if it seemed less fresh in later years, it was only because his style had entered the general bebop vocabulary.

One of Evans's musical trademarks was his left-hand harmonic reinforcement of his melodies. His close-to-the-keyboard hand positions enabled him to play near-legato repeated chords, even without using the sustaining pedal. The effect is similar to the locked-hands device, but because the right hand plays only single notes the line has all the range and fluidity of piano solo melodies generally. This effect dominates his final climactic chorus of *Autumn Leaves.*[28] The chorus from which this example is drawn begins with a well-wrought phrase on the chords Cmi^7-F^7-B$^\flat$:

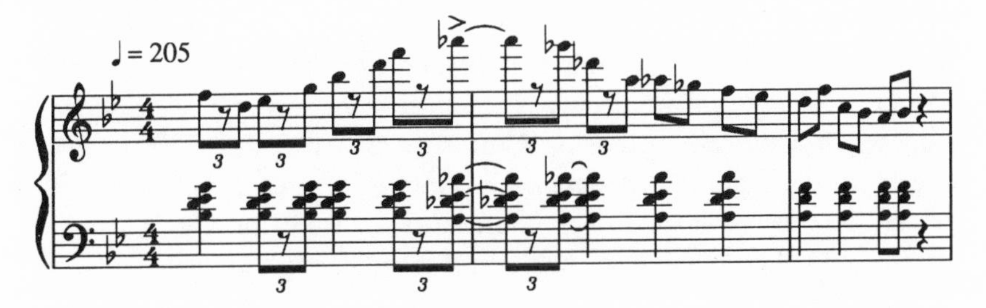

Twenty measures later he repeats this phrase, but with the eighth notes changed into one of his typical rhythms—♩ ♩ ♪♩ ♩ (triplet 3) ♩.

In September 1956, Evans recorded his first album as leader, Riverside 223. In it he established his career-long standard ensemble, a piano trio. He also presented three of his favorite pieces: *Five,* his humorous *I Got Rhythm* contrafact that he often used as a set-closer in club appearances; *Waltz for Debbie,* perhaps his most famous composition; and *My Romance,* the lovely Rodgers and Hart ballad. He plays *Waltz for Debbie, My Romance,* and parts of two other pieces unaccompanied, which was another career-long penchant. Aside from an excellent solo on *I Love You,* the album is good but not extraordinary. The trio sounds much like a typical bebop trio, with the bassist and drummer playing in a standard swing much of the time and Evans still revealing some influences from Bud Powell.

The following month he made an important recording as a sideman with George Russell: *Concerto for Billy the Kid,* which Russell wrote for him.[29] In it he has two solos; the second, built on the chords of *I'll Remember April,* he plays almost entirely without left-hand accompaniment, as though he were a saxophonist. Stylistically it stems from the melodic ideas of Lee Konitz and Lennie Tristano.

Evans's most important recordings as a sideman date from 1958 and 1959, when he was with Miles Davis's classic sextet. In this group he handled everything from the fast and aggressive *Oleo* to the lovely *Blue in Green* and *Flamenco Sketches* with equal mastery and artistry.[30] Yet in his mind he was the inferior player in the group, and this sense of inferiority, exacerbated by Davis's sometimes cruel humor, was partly responsible for him leaving the band in the fall of 1958, after serving as Davis's pianist for just a few months. Fortunately, he returned to play two sessions with Davis's group in early 1959, recording *Blue in Green, Flamenco Sketches,* and two other great pieces for the *Kind of Blue* album.

Evans provides the musical glue that holds together the unusual *Blue in Green.* He plays the introduction to his 10-measure theme.[31] Then he accompanies Davis beautifully and gently for one theme chorus. Evans's solo follows immediately, and in it he doubles the rate of harmonic change, creating two 5-measure choruses. Next he accompanies John Coltrane, still using 5-measure choruses. Then Evans plays a second solo, doubling the rate again for two 2½-measure choruses. He gracefully reverts to the 10-measure pace in time for Davis's theme chorus, and then plays two concluding out-of-tempo choruses.

Late in 1959 Evans re-recorded *Blue in Green* on a trio album in which he, bassist Scott LaFaro, and drummer Paul Motian create some gems of ensemble jazz (see pp. 208–10).[32] Again, he presents his theme at three different rates of speed. But this time the effect is more dramatic. There is first a crescendo of intensity as he moves from 10- to 5- to

2½-measure choruses, then a decrescendo of intensity as he moves back through 5- and then 10-measure choruses to the final peaceful coda. The entire performance is a miniature masterpiece, filled with harmonic, melodic, and contrapuntal ingenuity.

Evans's exquisite and often poignant rendering of the ballads in this album—they include *When I Fall in Love* and *Spring Is Here* as well as *Blue in Green*—is a joy. His beautiful, singing tone in the top voice of his chords would suggest a heightened sensitivity to the songs' lyrics; Lester Young and others have spoken about the importance of knowing the lyrics before interpreting a song properly. Surprisingly, however, Evans took a different approach: "I never listen to lyrics. I'm seldom conscious of them at all. The vocalist might as well be a horn as far as I'm concerned" (Lyons 1983: 224).

For Evans, harmony was an extremely important musical element. His harmonic palette was rich with raised and lowered fifths and ninths. He often reharmonized standard tunes, improving them dramatically in the process. When he composed he used many unconventional progressions. His *Fun Ride* is a 58-measure theme filled with unusually busy and chromatic progressions,[33] and his solo is an apparently effortless dance through this harmonic minefield.

Perhaps more than any other jazz pianist, Evans favored pieces in triple meter. Pianist Joanne Grauer, during a tribute concert to Evans, suggested that he, not Wayne King (the commercial band leader of the 1930s and 1940s), should be dubbed the "waltz king." The list of triple-meter recordings begins with his composition *Waltz for Debbie* (in which he played the theme in a rubato $\frac{3}{4}$ but the solo choruses in $\frac{4}{4}$), and includes *Someday My Prince Will Come, Elsa, Alice in Wonderland, Very Early, My Man's Gone Now, Love Theme from Spartacus, How My Heart Sings, B Minor Waltz, Sometime Ago,* and about two dozen others. He also composed *34 Skidoo* and *Let's Go Back to the Waltz,* pieces that move in and out of $\frac{3}{4}$.

Evans was a complex man. Most obviously, according to his music, he was a serious, sensitive, introspective, romantic human being. But there was a lighter side, too. For one thing, he enjoyed anagrams; two of his compositions were *Re: Person I Knew*—an anagram of his record producer during the 1950s, Orrin Keepnews—and *N.Y.C.'s No Lark*—written upon the early death of pianist Sonny Clark. His sense of humor emerges in the rhythmic games of his compositions *Five* and *34 Skidoo* and in the titles *Loose Bloose* (a blues), *Funkallero, Fudgesickle Built for Two* (which begins and ends as a fugue), and *Are You All the Things* (an *All the Things You Are* contrafact). His impish side bursts forth in rare brief moments. One occurs at the end of a recording session with Stan Getz and Elvin Jones, when he strides his way through an outlandish parody of the Gene Krupa Trio recording, *Dark Eyes.* Another is the hilarious coda to his overdubbed duet on *Little Lulu.*[34] There was also a distressed side, unfortunately, exacerbated by the drugs that affected many fine musicians. That side, rarely manifest in the recordings, is all too obvious in a posthumous solo album that never should have been issued.

Few younger pianists avoided Evans's impact. Either his touch, his contrapuntal interplay with bassists, his voicings, his melodic ideas, or some combination of these stylistic factors found their way into the playing styles of most pianists who followed him. He was bebop's most influential pianist.

Several other pianists played a vital role during the bebop expansion in the 1960s. Among them none was more important than McCoy Tyner (born in 1938), John Coltrane's pianist during 1960–65. Tyner represents the antithesis of Bill Evans. Evans caressed the keys to obtain the most lyrical sounds possible; Tyner pummels the keys and gets a harsh, dramatic sound. Evans often used legato phrasing; Tyner's preference for lengthy passages of detached sixteenths, played with an unusually high finger action, has been one of his most distinctive traits since the 1970s. Evans seldom rose above a *mezzo forte;* Tyner seldom drops that low in volume. Evans made the piano sing tenderly; Tyner makes it roar defiantly.

The two men employed largely different harmonic and melodic vocabularies. While Evans favored clear progressions based on tertian harmonies, Tyner favors harmonically ambiguous fourth chords. And while Evans's melodies clearly derive from the explicit progressions in his left hand, Tyner's melodies tend to wander through groups of four and five notes that he selects without obvious regard for the expected chord progressions. For example, in the last four measures of a blues in B$^{\flat}$, the standard chords are these—|CMI^{7} |F^{7} |B$^{\flat}$G^{7}|CMI^{7}F^{7}|—but in the third chorus of *Blues on the Corner,* Tyner and bassist Ron Carter play as follows:[35]

Carter begins the expected progression, but perhaps hearing Tyner's melody centered on F, shifts quickly to F and into the folk blues progression—F^7-$E^{\flat 7}$-$B^\flat$—and then on into a standard turnaround. Meanwhile, Tyner plays the $C{\rm MI}^7$-F^7 progression in his left hand, but in his right hand he plays some pentatonic figures that contradict that progression. The contradiction dominates the remainder of the chorus. He does end his melody on the tonic ($B^\flat$), but arrives at it unconventionally. The fourth chords in the left hand are ambiguous. For example, the chord G-C-F in the third measure is perhaps a $B^\flat$ 6/9 chord with a missing root and third, but also could be a G^7SUS, a CSUS, an $E^\flat$ 6/9 with a missing root, or some other chord.

Herbie Hancock (born in 1940) found some of Tyner's musical elements attractive and used them. He also listened to Bill Evans, whose chord voicings, touch, and left-hand-right-hand balance he used. His first recordings, with Donald Byrd and others, show that he was listening to Red Garland, Wynton Kelly, and Les McCann as well. Soon he synthesized these elements into an extremely versatile style. He was equally at home with middle-of-the-road bebop (*Nai Nai*), gospel- and folk-blues-tinged pieces (*Hush* and *Pentecostal Feeling*), and dissonant, harmonically adventuresome pieces that stretched the expected bebop limits (*Shangri-La* and *Free Form*).[36] This versatility has characterized his music for the past 30 years. Early in his career he composed and recorded the 16-measure folk blues *Watermelon Man,* which quickly became a "funky" jazz standard, and helped lay the foundation for his fusion-jazz successes in the 1970s and his rock successes in the 1980s. During his years with Miles Davis (1963–68, plus periodic reunions afterward) he played both the traditional bebop and the boundary-stretching pieces of that historic group.

Much of that boundary stretching was harmonic. While he contributed few new chordal *sonorities* to the vocabulary of jazz, he made important changes in chordal *emphasis.* That is, in some compositions he avoided the heavy reliance on ii-V-I progressions that characterized nearly all mainstream bebop. For example, in his piece *Maiden Voyage,* he uses only four chords: AMI^{7}/D and CMI^{7}/F in the *a* sections, B$^{\flat}$MI^{7}/E$^{\flat}$ and C$^{\#}$MI^{13} in the *b* section. Thus, the roots move by seconds and thirds, not by fourths and fifths, and the third omitted from the first three chords creates a harmonic ambiguity. Elsewhere Hancock puts a strong emphasis on enriched major triads and dominant sevenths. In *Speak like a Child,*[37] he voices the G$^{\flat}$ chord this way during the theme chorus—

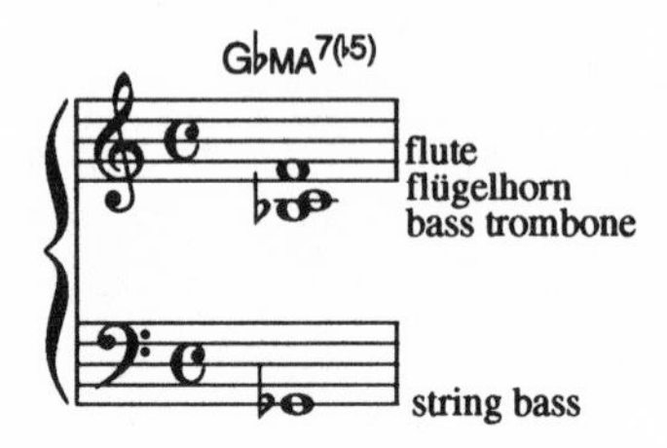

—and this way at one point in his improvisation—

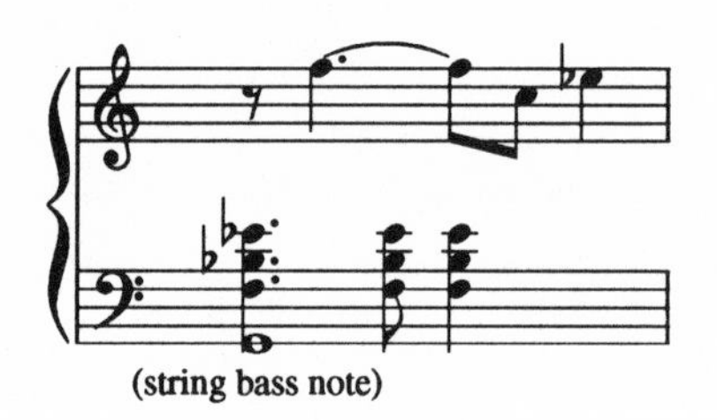

The piece begins with a rich pair of dominant seventh voicings:

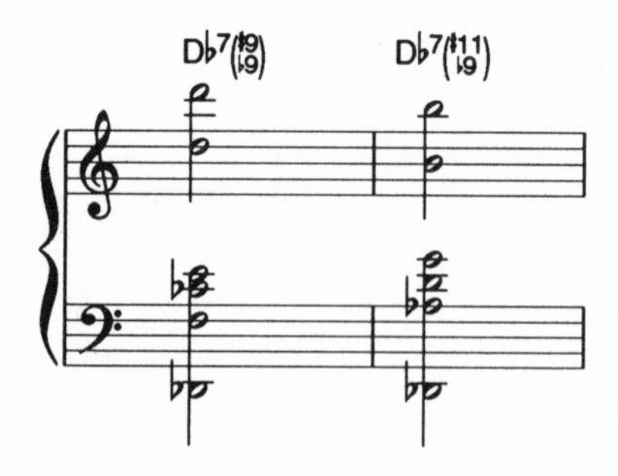

Dolphin Dance and other pieces have a similarly complex vocabulary.[38] As he explores the harmonic fabric of his compositions he commonly alters them. At one point in his solo in *Maiden Voyage* he changes the Cmi^{7}/F, which suggests the F Mixolydian scale (F-G-A-B♭-C-D-E♭-F), into something significantly different (F-G-A♭-B♭-C♭-D♭-E♭-F):

Just as Hank Jones and Tommy Flanagan have styles that are similar in many ways, so do Herbie Hancock and Chick Corea (born in 1941). Both can play a variety of jazz styles. Both were sidemen for Miles Davis in the 1960s (Corea replaced Hancock in 1968). Both became major figures in fusion jazz of the 1970s. Both continue to move between bebop bands and fusion bands from time to time. They have recorded and toured together, as duo pianists in 1978 and with their respective bands a decade later. And some of their compositions exhibit similarly varied harmonies.

Corea's style, however, is more dramatic and his tone quality is less lyrical than Hancock's. He favors a brighter sound in his chord voicings and a more percussive, detached articulation of melodic lines. His melodies derive more from McCoy Tyner and John Coltrane than from Bill Evans. In the blues *Matrix,* for example, his solo is an energetic, nonharmonic romp filled with melodies that weave in and out of the key. His aggressive, percussive style and his melodic chromaticisms are perfect for playing Thelonious Monk's compositions, as he proves ingeniously in *Trio Music.*[39]

Corea's improvisations and compositions regularly display a Spanish tinge, which stems from his association in the early 1960s with Mongo Santamaria and Willie Bobo. Two of his most famous compositions, *Spain* and *La Fiesta,* reflect that tinge. In his solo on Lenny White's *Guernica,* a piece utilizing the "Spanish Phrygian" scale, he displays a favorite melodic ornament—a short-note pair (shown in the box):[40]

Here the ornament appears in the right hand alone, but often he doubles the ornament two octaves lower in the left hand. In such moments this quick-note ornament stands out even more dramatically.

CHAPTER EIGHT

Bassists and Drummers

The heart of the rhythm section—and therefore of the entire ensemble—is the bassist-drummer rhythm team. The bassist is the most important voice in the harmonic structure that the band follows. And the two players provide jointly the basic pulse that guides the band. Since humans are not machines they cannot subdivide time with absolute precision. They cannot give all beats in a chorus of a piece, or even in a single measure, *exactly* the same length, though the differences are too small to be perceived consciously. So a drummer and bassist must find ways to synchronize their minutely unequal beats. They also must agree on where in time the bass notes fall in relation to the ride cymbal and high hat. These factors point to a unique relationship that ideally exists between drummer and bassist. They must be more than excellent players; they also must be compatible musically if the music is to swing. Drummers and bassists sometimes speak of their relationship as a marriage. It is an apt analogy.

Unfortunately, bassists and drummers were ill-served by the recording industry during the 1940s. Bass notes sound like unpitched quarter-note thumps in the worst of cases, and seldom have much presence even on the best recordings. Different tone qualities are almost indistinguishable, and the notes quickly fade into inaudibility, preventing the listener from hearing whether the bass line is legato or detached. The fade may have been deliberate; according to Red Callender, who began recording in the late 1930s, "the recording engineer would say 'Play short notes,' because it was hard to record the long bass notes. The equipment was very primitive."[1] Most of the drummers' cymbal sounds merged with the surface noise of 78-rpm records. Later, skillful remastering for LP and CD reissues improved the sound, but could not put back what the microphones did not capture. On record after record the high hat and bass drum are inaudible; only an occasional snare-drum fill comes through, and the ride cymbals strokes merge into a continuous shimmer.

Bassists

In the early 1940s, Jimmy Blanton (1918–42) of the Ellington band was the bassist best served by recording engineers; in fact, he was better served than was his rhythm partner, drummer Sonny Greer. The recordings captured much of his distinctive tone, which Gunther Schuller describes as "astonishingly full" and possessing a "firm clean-edged sound." He continues to describe Blanton's contribution:

> Most importantly, he was the first to develop the long tone in pizzicato. . . . Blanton . . . maximize[d] the natural resonance of the string by using as much of the fleshy length of the [right index] finger as possible—plucking the string with the finger parallel to the string, rather than plucking straight across at right angles; and [he] pluck[ed] the string at the point where it sets in vibration the maximum resonance. . . . Instead of the usual quick-decay of ordinary pizzicato playing, Blanton could produce whole notes or half notes or other longer durations at will. (Schuller 1989: 111)

Bassist John Clayton further suggested to me that to produce these longer notes Blanton must have held down each string as long as possible with his left hand before moving to the next note, a technique uncommon among other swing-style bassists.

Blanton was one of the first bassists to play solo lines that were significantly more than decorated bass lines (see Ch. 1, p. 7). These characteristics—his tone, his "legato pizzicato," his fluent solo lines—were inspirational for a whole generation of bebop bassists. He was not a bebopper—he died before the new style emerged—but he was the musical father of bebop bassists.

Oscar Pettiford (1922–60) was an early Blanton disciple. He was in a pioneering bebop group with Dizzy Gillespie in 1943–44, a group that made no recordings, unfortunately. During that period he did record with Coleman Hawkins, however, and impressed his colleagues with a half-chorus solo on *The Man I Love.*[2] The recording engineer undermiked him during the piano and tenor sax solos, so there is little to say about his walking bass lines. He is more audible during his solo, partly because the pianist and drummer drop to pianissimo. But even here the microphone is some distance from his instrument, and captures his between-phrase inhaling almost as loudly as his bass notes. Nonetheless it is clear, both from his breathing and from his theme-oriented solo, that he could play horn-like solo lines.

During the remainder of his short career Pettiford continued with his Blanton-inspired style, and even occupied his mentor's position in the Ellington band during 1945–47. As a sideman he played some excellent solos with Bud Powell, especially in *Blues in the Closet* and *Willow Weep for Me.*[3] He also led groups in the 1950s, and doubled occasionally on the "baby bass"—a cello tuned like a bass (in fourths) instead of like a cello (in fifths).[4] He contributed several fine tunes to the jazz repertory, most notably *Tricrotism, Bohemia After Dark,* and the blues *Swingin' Till the Girls Come Home.*

Charles Mingus (1922–79) was the same age as Pettiford, and, like Pettiford, also began his recording career in the mid-1940s. But the jazz world associates him more with the 1950s and 1960s, beginning with a well-known series of recordings with the Red Norvo Trio in 1950–51. He founded the Debut record company in 1951, and two years later issued pieces from the now-famous Massey Hall concert with Parker and Gillespie (see page 44). His most famous recordings as leader/-composer date from the mid-1950s and later (see pp. 228–29).

Mingus possessed the powerful, driving sound that Blanton pioneered; in some of his more exuberant solos he surpassed Blanton in power and energy, playing almost as if he were violently attacking his instrument. In his solos he used some unusual techniques; bending a pitch by using the guitar technique of pulling the string sideways, playing almost the same pitch alternately on two different strings, stopping and plucking a string simultaneously with the same hand (Priestly 1983: 62), and playing a rapid tremolo with two fingers on the same string.

His style, especially in the slow pieces, was ornate and often filled with irregular time values, as in *I Can't Get Started* and *Stormy Weather.* Still, he could reduce the complexity to get a folk-music effect, as in his dramatic *Haitian Fight Song* solo, or the New Orleans style, as in *Pussy Cat Dues.* His legato pizzicato and his powerful sound are well recorded in his album *Mingus Ah Um,* which contains *Pussy Cat Dues* and other classics.[5]

On those few early bebop records where the bass rings out with unmistakable authority and assurance, the bassist is usually Ray Brown (born in 1926). A native of Pittsburgh, he met Dizzy Gillespie on his first night in New York City in 1945, and began rehearsing with his group the following evening. The group included Charlie Parker, Max Roach, and Bud Powell—fast company for an unknown player from out of town! (Gillespie 1979: 230, 236.) He was in the sextet with Gillespie,

Parker, and Milt Jackson that traveled to Los Angeles in 1945, and in Gillespie's second big band during 1946–47. He played some solo passages on the big band's recordings of *One Bass Hit II* and *Two Bass Hit,*[6] but the solos are not as clearly recorded as his walking bass lines. In *Two Bass Hit* there are three blues choruses in D♭. Here Brown makes liberal use of passing tones (p) and chromatic neighboring tones (n) in an ingenious wedding of harmonic function and melodic interest (these are the last four measures of the first chorus):

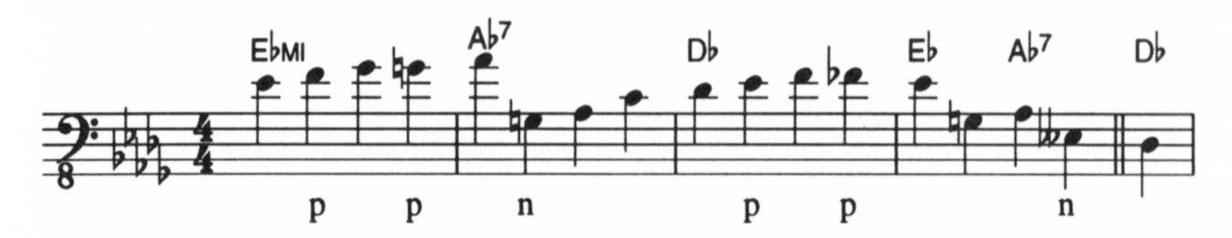

This is the type of line that the best young bassists were beginning to create. All that is absent from the recording is the legato pizzicato that characterized recorded bebop a decade later.

With an auspicious start in the bebop world Brown was soon an established master. He performed and recorded with Parker, Powell, Lester Young, Ella Fitzgerald (his wife during 1948–52) and others in the late 1940s and early 1950s. Most of these associations stemmed from his participation in Norman Granz's Jazz at the Philharmonic concerts. He was in that touring assemblage on 18 September 1949, when Oscar Peterson made his JATP debut in Carnegie Hall. The two men played some duets on that occasion, and a few months later, after Peterson joined the tour, he and Brown recorded their inaugural duet album (Verve 2046). The Peterson-Brown partnership was to become enormously productive and greatly admired, first as a duo, then as part of the Oscar Peterson trios of 1956–65 (see page 207).

The early recordings with Peterson fail to capture Brown's sound adequately. In most he is too far back in the mix to allow the ends of his notes to be heard. His aggressive playing, with an abundance of eighth-note fills, comes through well enough to identify him, however, and no other bassist generated such a forceful and relentless swing. The clearest recordings from the period are his two sessions with the Milt Jackson Quartet—the precursor of the Modern Jazz Quartet (see pp. 211–14).[7] In these recordings he plays a mix of legato and detached playing, both in quarter-note walking lines and half-note two-beat passages. His stylistic identifiers are clear: many cleanly articulated fills, especially in slow and medium tempos; a stinging attack with an unusually wide overtone range; and again, an unmistakable forward momentum.

Returning to the Peterson recordings, it appears that the preferred rhythm section sound during many of Peterson's solos in the early 1950s was a crisp, dry, detached guitar stroke matched with detached bass notes. Then, as recording technology improved, that sound seemed to change to detached guitar chords above a *legato* bass line. An early sampling of that sound dates from late 1953, with *Pompton Turnpike, Soft Winds,* and *Cherokee* (Clef 694). In the 1956 recording done in Stratford, Ontario (Verve 8024), the bass has such excellent presence that Brown is much better served than his colleagues. There are problems in this recording; the foot-tapping on a hardwood floor and Peterson's exuberant, involuntary nasal singing are distracting. Even so, in *How High the Moon* Brown plays two enjoyable quote-filled solo choruses—he played few solos during his years with Peterson—and the main body of the piece is a priceless aural textbook on how to play walking bass lines on that tune.

During the past decades Brown's sound and skill have remained undimmed. He is an agile, inventive, and often humorous soloist. His *arco* technique is excellent, though he seldom reveals it. But he shines most brightly as an accompanist. Examples of his beautiful bass lines are legion. Here are two contrasting types—two excerpts from an extremely slow piece and one from a medium-tempo piece—both from his album *The Red Hot Ray Brown Trio,* with pianist Gene Harris and drummer Mickey Roker.[8] The first two are from *Street of Dreams.* During the first chorus Brown plays "in two" (emphasizing beats 1 and 3 in the four-beat meter) while Harris plays the theme. But Brown does much more than play half-notes; he embellishes the long notes by preceding beats 1 and 3 with some shorter notes. These embellishments serve to generate a forward momentum into the strong beats. They also help the players and audience alike in keeping an accurate count of the strong beats, which are 2⅔ seconds apart.

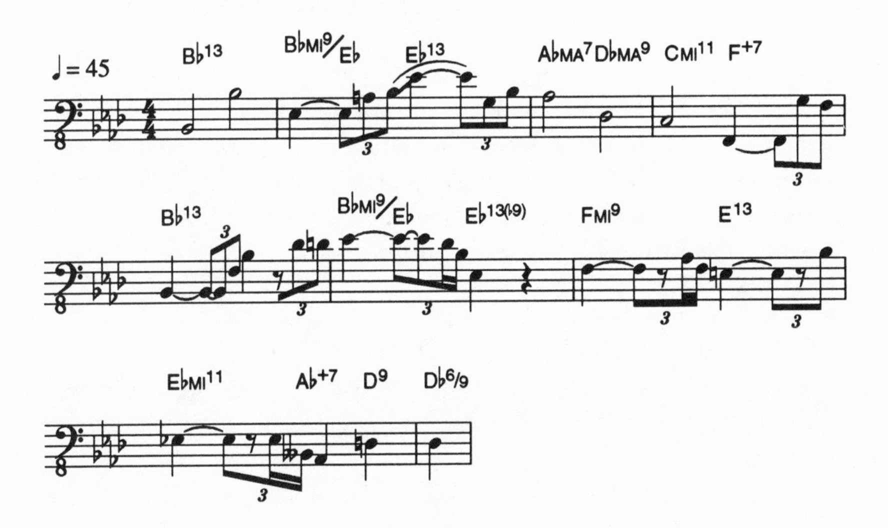

In the second chorus he plays "in four" (emphasizing each beat). Again, however, this is no mere string of quarter-notes, but a highly decorated line that supports and complements Harris's improvised chorus and adds a strong sense of swing to this slow piece. Notice the wide range traversed in measures 4 and 5.

In the third chorus, Brown and Roker move into double time (subdividing each beat and creating the effect of $\frac{8}{8}$ meter), during which Brown's lines are less ornate, but the swing sense is greater.

His bass lines in *How Could You Do a Thing like That to Me* are simpler than in *Street of Dreams.* The tempo is much faster, and does not provide the opportunity for large numbers of between-beat interpolations. Instead, Brown walks. And what a classy walk it is, especially since the *a* section of this *aaba* tune contains little more than repetitions of the two-measure progression I-vi-ii-V—four repetitions in the first *a,* three in each of the other *a* sections. Out of these ten statements of the progression, eight are clearly different, and no two are identical. In several he substitutes III for I (G^7 for $E^{\flat}$). Also, he interchanges different chord qualities on a common root; thus, CMI becomes C or C^7, FMI becomes F or F^7 in several places. Sometimes, in the interest of melodic continuity, he avoids roots altogether (as in measures 7, 9, and 10 below). The chorus clearly illustrates the principles of good bass line building that Brown helped establish in the late 1940s.

Percy Heath (born in 1923) began his professional career in the late 1940s, when he and his brother, saxophonist Jimmy Heath, moved from Philadelphia to New York City. He began recording in 1948, on sessions led by Kenny Clarke and Howard McGhee, and by the mid-1950s had recorded with nearly every major bebop musician. His inventive bass lines, such as in the multi-chorus *Walkin'* by Miles Davis[9]—a bassist's textbook on the blues—grace many historic recordings. He has had one primary association during his career—the Modern Jazz Quartet (pp. 211–14). As his 40-year tenure with the MJQ suggests, Heath is a superb ensemble player. His greatest strength is his ability to mesh with a drummer—mostly Connie Kay, of course, but others outside of the MJQ, as well—and to provide solid harmonic and rhythmic support for the ensemble. His solos tend to be less flashy or dramatic than Mingus's or Brown's. But his solos on "baby bass" with the Heath Brothers (originally Jimmy, Percy, and Tootie Heath plus pianist Stanley Cowell) are among the best on that instrument.[10]

In his mature years, Red Mitchell (1927–92) developed a highly unusual bass tone quality. During the 1950s, while functioning as a sideman for many different leaders, his sound was well within the mainstream of bebop bass sound except for a slight airiness. But in 1966 he switched to cello tuning on his bass (C-G-D-A, an octave below the cello, instead of the standard E-A-D-G). At the same time, he began adjusting the tone controls of his amplifier to create a soft, unfocused sound in the lowest notes and to emphasize the upper harmonics in higher notes. The result was an airy tone quality that sounded gentle, not muscular. This airy tone and his frequent habit of strumming the strings with his right thumb contributed greatly to his unusual style.

By the 1960s his solo melodies were more flexible rhythmically than those of his contemporaries, and were filled with more upper chordal elements and fewer of the roots and thirds that he often favored earlier. His solo in *The Seance,* a blues recorded with Hampton Hawes, is an example of this "un-bass-like" rhythmic and melodic freedom. Mitchell had an amazingly keen ear for pitch and dynamic subtleties. He could play pizzicato in a profoundly emotional way; his beautiful, moving solo in the slow, mournful waltz *Nature Boy,* by Art Pepper's quartet of 1979, is a stunning example.[11]

Paul Chambers (1935–69) was another prominent bebop bassist in the late 1950s and early 1960s. His primary forums were the Miles Davis groups of 1955–63, though he also performed and recorded with many other great bebop players during that period. And, unlike most bassists, he served as leader for seven albums. He had a softer, mellower tone, and played slightly less legato, than Ray Brown. His solos were less fiery but just as inventive as those by Pettiford, Mingus, or Brown. He played *arco* solos more often than most, usually with such rapid motion that his tone and intonation were hard to assess, though in *Yesterdays* and *The Theme,* from his album *Bass on Top,* he reveals a good tone and accurate intonation.[12]

Chambers's solos share much in common with those of wind players, pianists, and guitarists of the late 1940s. The characteristics of his solos include 1) eighth-note-dominated phrases that begin between beats one and two; 2) sets of triplet quarter-notes that begin unexpectedly on beats 2 or 4 instead of on beats 1 or 3; 3) the combination of Figure 4B and 4A (see p. 32) from Parker's vocabulary (a habit common to many bebop players of all instruments); 4) Figure 5B; and 5) Figure 3, also from Parker's vocabulary. These features appear as numbered in the example below, his second solo chorus in *Sweet Sapphire Blues* from a John Coltrane session:[13]

As is true of all bassists, Chambers's chief contribution was his accompaniments more than his solos. He was a great team player, in terms of both the swing he helped generate with drummers—especially Philly Joe Jones and Jimmy Cobb—and the harmonic/melodic foundation his well-crafted bass lines provided. To cite but one example from a legion of fine recordings, he is the hero of Art Pepper's recording of *Softly, as in a Morning Sunrise.*[14] Here he produces a definitive set of walking bass lines for this piece, and drives the quartet with an irresistible swing.

Sam Jones (1924–81) entered the upper echelon of bebop contemporaneously with Paul Chambers, but his career flourished twice as long. Both men were much in demand for their reliable time keeping and inventive bass lines. They shared some melodic habits; both used Figure 4A (Chapter 3), both played quarter-note triplets across beats 2 and 3 or 4 and 1, and both used this rhythm—|♩ ♫ 𝄾♩ ♩|—as a phrase ending. Just as Chambers had a great association with Miles Davis, Jones had an equally productive tenure—1956–65—with Cannonball Adderley. During several of his years with the Adderley brothers Jones's rhythm partner was drummer Louis Hayes. This exciting team was as vital to the musical success of the group as were the Adderleys themselves.

As a soloist Jones was usually more aggressive and agile than Chambers, and employed a tone and attack reminiscent of Charles Mingus. Yet at times he was the more conservative of the two soloists, for unlike Chambers, he often simply walked throughout his solo. His walking bass lines were more legato than Chambers's. They became extremely legato in the 1970s, when—perhaps under the influence of Ron Carter's style (see below)—his solo style also began to include an exaggerated vibrato on long notes. Jones did not share Chambers's enthusiasm for the bow. Conversely, Chambers did not share Jones's enthusiasm for the cello, which Jones used on several of his albums. Jones and Chambers both were ill-served by the recording engineers, or by the recording equipment, of the 1950s and 1960s. They are usually far in the background in the sound mix, even during their solos. Jones fared better in the 1970s, by which time, unfortunately, Chambers was dead.

Scott LaFaro (1936–61) began his recording career in August 1957, as a sideman for bebop clarinetist Buddy DeFranco (Verve 8383). Four months later he played on pianist/vibist Victor Feldman's first trio album made in the United States (Contemporary 7549).[15] Both his walking bass lines and his solos were powerful, agile, and legato, much in the manner of Ray Brown. His horn-like solo lines, sprinkled with several Parker figures, were solidly within the bebop vocabulary. It seemed that another first-rate conventional bassist had emerged to continue the tradition. Soon, however, he began to rethink the role of bebop bassists, and became a major innovator on his instrument.

The main forum for his innovations was the Bill Evans Trio, with drummer Paul Motian, during the years 1958–61. In *Autumn Leaves* from their first album together, he engages in a dialogue with Evans instead of playing a normal solo.[16] In much of *Witchcraft* from the same album, LaFaro interacts rhythmically with Evans rather than playing the usual walking bass lines. But the change is incomplete in this album, for on other tunes he plays conservative, even dull, two-beat accompaniments.

In succeeding albums LaFaro showed that the move away from standard walking bass lines previewed in *Witchcraft* was not momentary, but becoming the norm. In many medium-tempo pieces he plays rhythmically varied lines to accompany the first two or more choruses by

Evans (sometimes using the ♩ ♩ ♫ rhythm that Evans also favored), then breaks into walking lines. In some pieces he never does walk. His masterpieces of rhythmically varied bass lines date from the afternoon and evening of 25 June 1961, when the trio recorded in concert at New York City's Village Vanguard.[17] The musicians were in top form, engineer Dave Jones captured their playing splendidly, and the result was some beautifully subtle and intricate trio jazz (see pp. 208–9).

In *Solar*, LaFaro presents a textbook on the new role of bebop bass playing that he perfected. Throughout the nine-minute performance he functions as a melody player, providing rhythmically independent counterpoint to Evans's own octave-doubled improvisations. In the opening choruses his lines are variants of the theme; afterward he is as busy improving his melodies as is Evans. The example below shows the thematic pick-up notes leading into the fifth chorus and next two choruses. In the fifth chorus he develops a three-beat figure much as he might in a bass solo; but he is near the beginning of *Evans's* solo, not his own. At the end of this chorus he leads us to expect the standard walking bass line, but then denies the expectation.

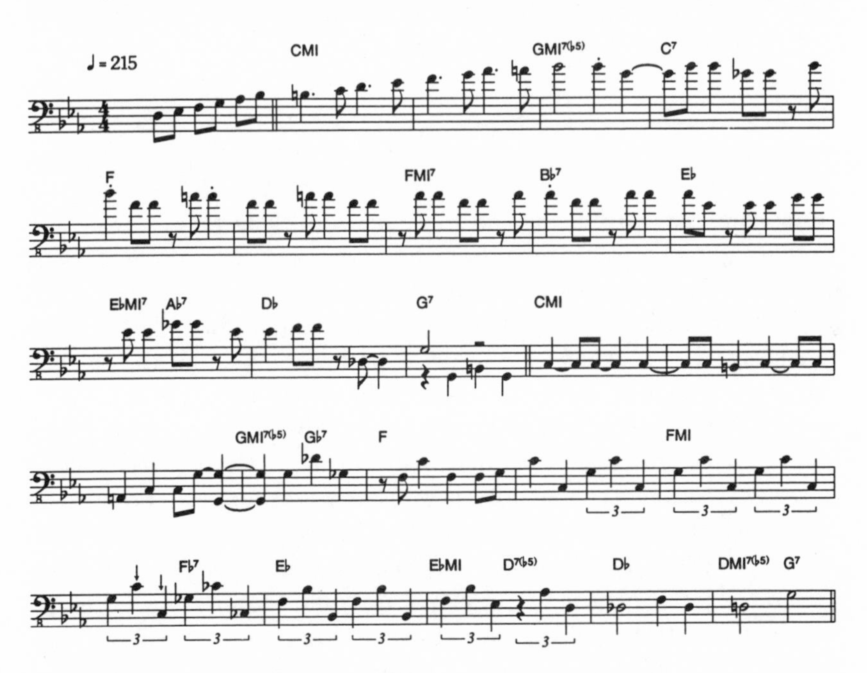

His playing throughout the piece is similar to this example; it differs markedly from the styles of Ray Brown, Paul Chambers, and other contemporaries. It has some of the rhythmic freedom explored by bassists in the action-jazz style during the late 1950s and 1960s. LaFaro, in fact, had some professional associations with that movement, the most famous being his participation in Ornette Coleman's album *Free Jazz.*[18]

Tragically, LaFaro died in an automobile accident eleven days after making those remarkable trio recordings. Apart from some poorly recorded pieces done with Stan Getz at the Newport Jazz Festival three days before the accident, the Village Vanguard recordings with Evans are his last. His legacy of rapid and lightly played melodic bass lines lives on, however, perpetuated by his successors in Evans's trios—principally Chuck Israels, Eddie Gomez, and Marc Johnson—and by many other players. He made a substantial contribution to the language in his short recording career.

Ron Carter (born in 1937) is another distinctive bass stylist in bebop. As with most major players, Carter put his style together gradually over several years. When he began recording in 1960, he seemed to have the standard deep, rich tone and legato articulation that most of his older colleagues had developed earlier. Soon after he joined Miles Davis in 1963, he began adding to his playing a conspicuous portamento, especially in medium and slow tempos, using it on long notes and sliding up and down through a wide pitch range:[19]

Portamento embellishments and those beautiful long tones remain the most readily identifiable features of his style, ones that appear in both his accompanying lines and his solos.

Carter studied the cello for about seven years before taking up the bass at age 17. Throughout his career he has doubled on the smaller instrument, using the standard cello tuning (C-G-D-A) instead of the tuning (E-A-D-G) used by most bassists doubling on cello. Also, when fusion jazz became popular in the 1970s, he began playing electric bass on many recording sessions. But it is his stunning pizzicato on upright bass for which he is most famous. An excellent sampling is on his album, *Third Plane.*[20] The album, done by the famous Miles Davis rhythm section of Herbie Hancock, Tony Williams, and Carter, captures every nuance of his style. It is a treasure trove of pizzicato techniques: portamento effects, vibrato, double stops (including parallel octaves), harmonics, the most legato pizzicato lines in jazz, and those beautiful long notes on the lowest (E) string—notes whose overtone mix evolves continuously as the notes ring. (The final note of *Lawra,* which lasts about seven seconds, is a particularly fine example in this album.) Occasionally some unusually low note sounds, for Carter has an extension attached to the instrument that allows him to play down to low C on his E string (John Heard and Rufus Reid also use the C extension).

LaFaro, Carter, and other bassists exemplify the amazing evolution of bass technique that has occurred since 1940. Most early bebop bassists were time-keepers primarily. In the 1950s, Pettiford, Mingus, and Brown, building upon the foundation Blanton provided, pushed on to higher levels of virtuosity. LaFaro and Carter went further, and joining them were other virtuosi, such as Niels-Henning Ørsted Pedersen, Eddie Gomez, Stanley Clarke, George Mraz, and John Clayton. I have only scratched the surface in this brief survey. Todd Coolman, in his fine anthology of transcribed bass solos (1985), discusses and musically illustrates the styles of fifteen bebop bassists and lists nearly 150 more.

Drummers

During the first decades of jazz history the drum set, and the techniques of playing it, underwent several fundamental changes. By the 1940s the typical drum set consisted of bass drum, snare drum, one to three tom-toms, high hat, and one to three suspended cymbals of various diameters and thicknesses. This basic set has undergone only minor alterations during the past 50 years. Thus, the emergence of bebop drumming coincided with the emergence of the modern drum set.[21]

The word "drummer" is inadequate to describe the players, just as "drums" is inadequate to describe the array of instruments under consideration here. The older term "trap set player" is obsolete. Some players prefer "multi-percussionist" or "multiple percussionist," which are more accurate but probably too lengthy to gain widespread acceptance. Further, they raise the potential confusion with the term "percussionist," which in jazz commonly refers to players of miscellaneous instruments not included in the typical drum set. "Drummer-cymbalist," while more accurate than "drummer," would likely prove to be as unpopular as the other polysyllabic terms. So "drummer" it is.

During the 1930s, drummers typically backed ensembles and soloists with a steady quarter-note stream on the bass drum and a slightly more complex pattern on the snare drum or cymbals. Extra cymbal or drum punctuations usually concluded four- and eight-measure phrases. These drummers sometimes reinforced the distinctive rhythms of the arrangements (such as the brass chordal punctuations), but often ignored those rhythms and instead played favorite brief fills in predictable places.

Different drummers had their individual ways of playing background patterns and end-of-phrase fills. Jo Jones (1911–85), the master drummer with Count Basie's band during 1936–48, probably had the greatest impact on early bebop drummers. One of his habits was to play the standard high-hat pattern—— and to separate the cymbals just partially on beats 1 and 3. The result was a sustained, ringing sound resembling that of a sizzle cymbal (Brown, T. D. 1976: 445). He also played several variants of the basic pattern. More significant, he often played these patterns on the ride cymbal, thereby producing the more legato sound that became a hallmark of bebop drumming (Brown 447–48). Jones, unlike other prominent swing drummers, avoided using tom-toms, cowbell, and woodblock during solo passages, using instead only snare and bass drums, and high hat and crash cymbals (Brown 449). (Bebop drummers followed his lead, perhaps because the sharp, dry sounds of woodblock and cowbell and the insistent "jungle drumming" of tom-toms—à la Gene Krupa in *Sing, Sing, Sing*—reminded them of the style they wished to leave behind.) On the other hand, he used his bass drum as an independent voice in his rhythmically polyphonic solos, something rare among swing drummers but soon to be common among bebop drummers (Brown 452–53).

Jones's phrasing in his solos is distinctively different from his colleagues' phrasing. Instead of using the normal, predictable two- and four-measure phrases, "he often overlaps rhythmic ideas from one section to the next" (Brown 453). Finally, his subtle virtuosity provided a standard of excellence for which his successors strove. His brush technique was singularly masterful, as was his stick and cymbal technique on the high hat.

According to Brown, "bop drumming differed from all previous jazz drumming styles in three ways: the first was the consistent use of the ride cymbal to create a wash of sound within the ensemble; second was the gradual removal of the bass drum from its time-keeping role; and third, the evolution of coordinated independence" (Brown 463). Each of these points deserves further amplification.

The shift to the ride cymbal was a gradual one that began with Jo Jones, Dave Tough, and others in the 1930s; it was a common practice by the mid-1940s. These cymbals vary widely in acoustical properties, depending on subtle variations of construction. Drummers are keenly sensitive to these varied cymbal sounds; "the selection of a ride cymbal has . . . become as important to the jazz drummer as the mouthpiece and reed selection has always been to the brass and woodwind players" (Brown 465–66). Drummers regularly leave their drums and hardware in a jazz club from night to night, but they take their prized cymbals home each night.[22]

Specifics of bass drum playing are difficult to discern even in recent digital recordings, let alone recordings of the 1930s and 1940s. But swing drummers often played heavily and continuously on beats 1 and 3 or on all four beats. Sometimes they nearly obliterated the string bass notes. During solo breaks Jo Jones and others used it in more varied ways, but the swing norm was thump!—thump!—thump!—thump! Kenny Clarke was perhaps the first to use the bass drum for off-beat punctuations while backing the ensemble and other soloists (see Chapter 1), though Buddy Rich did the same thing in 1939, while playing in the Artie Shaw band.[23] Eventually most bebop drummers began playing all four beats softly except when playing intermittent punctuations, or "bombs." On recordings only the punctuations come through; the four-beat bass drum tapping is primarily for the drummer's benefit, analogous to foot-tapping by other instrumentalists.

"Coordinated independence" refers to the four different rhythmic layers that bebop drummers created on the ride cymbal, high hat, bass drum, and snare drum. This change also happened gradually and began before bebop existed. The biggest change was not in the emergence of four sound layers, but in the timing of the snare and bass drum layers. Swing players frequently marked off four- or eight-measure phrases with simple and largely predictable drum punctuations, ignoring the accents of the ensemble or solo melodies. Bebop drummers tended to adapt their punctuations to the melodic lines of the ensemble and soloists. If rhythmic accents seemed called for they would play them no matter where they fell, and no matter how long or short, symmetrical or asymmetrical were the resulting phrase lengths. For example, in the Gillespie big-band recording of *Our Delight,* Kenny Clarke reinforces most of the off-beat *tutti* chords, and plays fills that lead into many of those off-beat chords.

Since Kenny Clarke (1914–85) is the undisputed founding father of bebop drumming, it is a shame that so little of his pioneering work exists on record. He began recording in 1938, but other than a brief moment with Sidney Bechet in 1940 and the jam sessions at Minton's in 1941 (see pp. 9–10), his early recordings seem solidly within the swing idiom. The American Federation of Musicians strike of 1942–43 and then military service interrupted his recording activities until 1946. Once he resumed recording the engineers usually placed him away from the microphones, and failed to capture clearly his splashy-sounding ride cymbal. Only the strokes on his partly opened high hat, which he used to accompany piano solos, come through clearly on his early bebop recordings.

In the early 1950s recording technology was kinder to him at times. These recordings reveal that in his ride cymbal playing he placed a heavy emphasis on each beat, and played the between-beat notes with a light stroke: . They also suggest that his creativity was governed by the rhythmic features of the piece at hand and the energy of the other players. Lackluster themes and solos elicited similar drumming from him; and he seemed uninspired by ballads, responding to them with simple, plodding brush strokes.

In 1954 he began recording at Rudy Van Gelder's studio, and for the first time his set came alive on record. Van Gelder miked the drum set closely and captured the cymbal strokes cleanly and clearly. One happy result is the famous Miles Davis recording of *Walkin'*.[24] The clarity of the recording allows us to appreciate why drummers take pride in their cymbals, for Clarke's ride cymbal sound is beautiful here. His playing during the last six choruses—two by Davis, two of a secondary theme, and two of the main theme—illustrate the beauties of bebop drumming at its best, even though it is marred by a small rhythmic stumble (by bassist Percy Heath?).

In the mid-1950s, recordings made in Van Gelder's living room/recording studio reveal the intricacies of Clarke's best brush work in ballads, such as Milt Jackson's recording of *I've Lost Your Love.*[25] They also show that he used his bass drum sparingly as a bomb-dropper, and that he played relatively few punctuating fills on the snare drum. They show that he often adjusted the rhythm of the ride cymbal pattern to this: . Finally, they show him to be at his best as an accompanist. His solos sound as though he sometimes froze up, unable to move as fast as he would have liked.

By the 1960s he had changed his style in some ways. When playing for a powerful soloist such as Dexter Gordon he played a much more aggressive and animated accompaniment than he had in the 1950s. Perhaps by then he had gleaned some ideas from younger drummers such as Philly Joe Jones or Elvin Jones, or perhaps he developed this change independently. A good sampling of his 1960s style is Dexter Gordon's album *Our Man in Paris,* especially the piece *Broadway.*[26] Throughout the 1960s and early 1970s he co-led a big band in Europe with pianist Francy Boland. This ensemble of European and American players was one of the world's finest big bands, largely because of Clarke's masterful and propulsive playing.

The earliest recordings by Max Roach (born in 1925), done with Coleman Hawkins's group and with Benny Carter's big band during the years 1943–45, show him to be an excellent swing-style player. During that time he was also a member of the unrecorded bebop band co-led by Gillespie and Pettiford. His record debut as a bebop drummer, the famous Parker session in November 1945 (pp. 17–19), established his importance in the new idiom. For the rest of the decade he was considered one of the top bebop drummers in the world.

Beginning in the late 1940s his recordings showed him to be a more aggressive player than Clarke. Where Clarke's bass drum bombs occurred every few measures, Roach's fall every two to four beats:[27]

Where Clarke played just an occasional snare drum fill to supplement his ride-cymbal pattern, Roach played so many that his snare drum often was more active than his cymbal:[28]

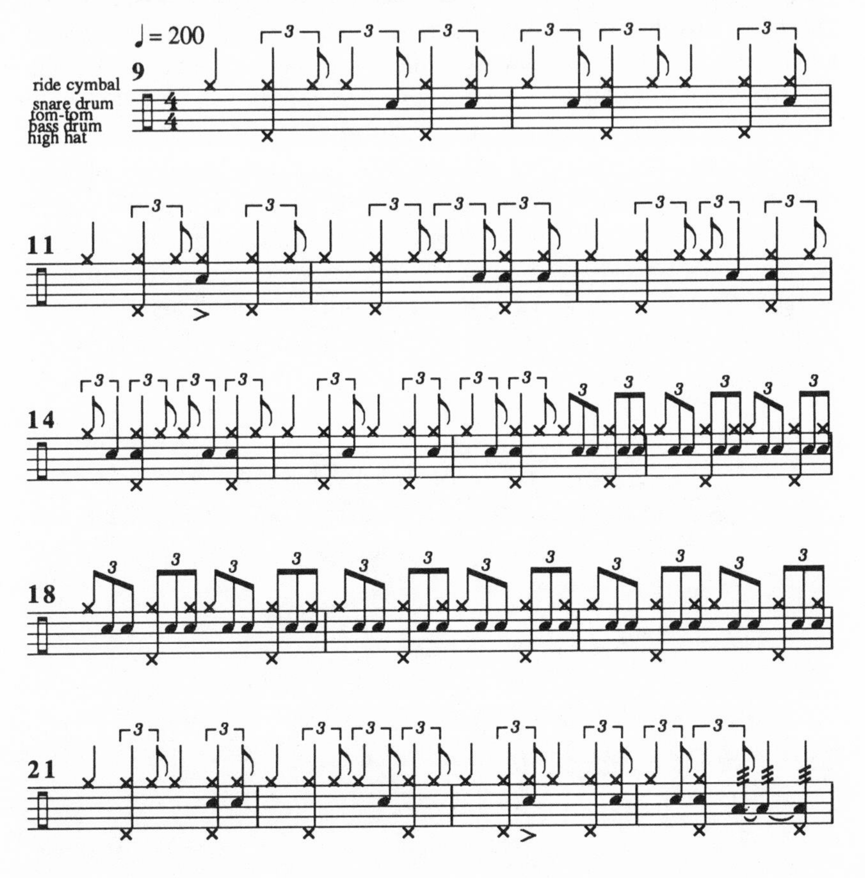

Roach's ride cymbal sounded different from Clarke's, partly because its tone quality was clearer and more bell-like, and partly because of a different accentuation pattern: (compare with Clarke's, above).

The most dramatic difference between these two bebop pioneers was in their respective solos. Roach soloed far more frequently, both as a sideman and as a leader or co-leader, than Clarke did. Musicians use the term "melodic drummer" to describe someone who develops rhythmic ideas throughout a solo instead of simply showing off technique. In that sense, Roach is a supremely melodic drummer; his solo in *Stompin' at the Savoy* is a striking case in point.[29] He often starts his solos with simple patterns and gradually increases the complexity, as in Parker's *Cosmic Rays.*[30] He is a master of motivic developments and sometimes uses rhythmic motives drawn from the theme of a piece. He also plays solo pieces, including, since the late 1950s, solo pieces in asymmetric meters.

Un poco loco, with Bud Powell (see pp. 205–6) is a justly famous recording. In it, Roach's ingenious accompanimental ostinato garners more attention than does his solo, though his solo is effective as well. Several fine examples of his accompanying and soloing date from the quintet dates that he co-led with Clifford Brown. Among them are *Clifford's Axe, Daahoud* (2 takes), *Blues Walk* (2 takes), and *Flossie Lou,* illustrated above.[31] During the years with Brown he would play one or more choruses of trading fours or eights with the wind players and then one or more solo choruses. He followed this solo procedure in subsequent years, as well. Some of his favorite patterns are in the example below; sometimes he plays the eighth notes evenly, other times as or as :

His most famous lick is this four-measure phrase upon which he bases an unaccompanied solo, *Big Sid,* named in honor of the great swing drummer Sid Catlett:

He is also famous for his tribute to Jo Jones, *Mr. Hi Hat,* a short, dazzling piece that he plays on the high hat alone.

In 1989, Roach and Dizzy Gillespie gave a duo concert in Paris. The recording of their 90-minute concert (A&M CD 6404) captures Roach's sound with crystal clarity, and reveals every detail of his accompanying and soloing habits. The improvement in fidelity between his recordings of the 1940s and this unusual concert recording is astounding. So is the difference in musicality, for Roach's artistry at age 63 was far higher than when he first became a leading figure in bebop.[32]

The colorful Art Blakey (1919–90) was another of the pioneering bebop drummers. In the early 1940s he played in Fletcher Henderson's swing-style big band. Soon he established his bebop credentials, in Billy Eckstine's early bebop band and on three of Thelonious Monk's early recording sessions. In the early 1950s he played with Charlie Parker and Miles Davis, and then began leading groups himself, including a quintet that included Clifford Brown for a short time (see pp. 128, 219–20). He used the name Jazz Messengers for his groups, a name he had used for some short-lived ensembles in the late 1940s. In the 1950s the Messengers were a quintet, in the 1960s a sextet, and in the final years a septet.

Blakey's two most idiosyncratic devices are his press roll and his shuffle beat. When he began that great press-roll crescendo in the last measure of a chorus he seemed to lift the band and the audience in his powerful arms and deposit them on the threshold of the next chorus. We can notate the basic features of his shuffle beat—showing the subdivided beats alternating between the ride cymbal and the snare drum—but not its forward momentum. The measures before and after the end of the first chorus of *Moanin'* illustrate both of his musical signatures:[33]

Blakey was neither the subtlest nor the most refined of bebop drummers. His single-stroke rolls, for example, were often energetic but imprecise (some would say sloppy), especially early in his career. But, like pianist Horace Silver, he had the technique he needed to execute his dramatic musical concepts. Indeed, the specifics of the notes and phrases he played were much less important than his overall approach to music. As a leader of the many different Jazz Messengers during 1955–90 he exerted an uncommon amount of control over the music, a subject discussed further in Chapter 10.

Some early bebop drummers, though they developed less distinctive styles than Blakey's, made important contributions to the music by their consistency and thorough mastery of the new idiom. Shelly Manne (1920–84) was one of these tradition bearers. He took part in some of the earliest bebop recordings, and later became associated with "West Coast jazz," though he was a New Yorker and did not move to Los Angeles until he was in his thirties. West Coast jazz usually denoted quiet, restrained, even anemic jazz. To be sure, he could be the most unobtrusive of brush-wielding drummers, able to keep a low enough profile to satisfy even the minimalist demands of Lennie Tristano (see page 218). But he also could be a dynamic and inspirational player, as many recordings done under his leadership prove (see page 218). Like Max Roach, Manne was skilled in "melodic drumming," particularly with mallets, such as in the appropriately titled *Mallets.*[34] And he was ingenious in transferring Afro-Caribbean rhythms to the drum set (for example, *Viva Zapata No. 1,* with Howard Rumsey's Lighthouse All-Stars.)[35]

Roy Haynes (born in 1926) was a professional swing and dixieland player as early as 1944. He quickly acquired the new vocabulary when he went to New York and began working on 52nd Street, and by 1948 the recordings he made with Lester Young show his completed conversion to bebop. Thereafter he played with the cream of the bebop world—Bud Powell, Miles Davis, Charlie Parker, Sarah Vaughan, Thelonious Monk, and others. In the early 1960s he walked the line between bebop and action jazz during occasional engagements with Coltrane and Eric Dolphy. He moved into fusion in the late 1960s and 1970s. Largely a freelance player, Haynes seldom plays with any one group for an extended period.

Joe Morello (born in 1928) is best known for his work with the Dave Brubeck Quartet during 1956–67. He was with Brubeck for the premiere recordings of many polymetric pieces and pieces in asymmetric meters (such as $\frac{5}{4}$ and $\frac{7}{4}$), works that formed a large portion of the Quartet's repertoire beginning in 1959. Thus, he pioneered in solving some rhythmic challenges that were either extremely rare or previously unknown in jazz.[36] Since leaving Brubeck his primary activities have been as a drum teacher, though he has performed publicly for brief periods.

Most of the early drummers discussed so far entered the profession as swing-style drummers, then adopted, and helped formulate, the new idiom during the 1940s. But two world-famous bebop drummers remained in the swing idiom until the 1950s: Buddy Rich (1917–87) and Louie Bellson (born in 1924). In terms of dazzling technique, they are to the drum set what Art Tatum is to the piano. Like fine athletes, they quickly learned how to achieve maximum effect with minimum effort (though Rich's profuse sweating appeared to contradict this fact). Like entertainers—both were tap dancers as children and Rich spent his childhood working in vaudeville—they believed in giving the audience a good show. Their solos are visually and aurally spectacular. Perhaps because of their early careers as big-band drummers—Rich with Bunny Berigan, Harry James, Artie Shaw, Benny Carter, and Tommy Dorsey; Bellson with Benny Goodman, Dorsey, James, Count Basie, and Duke Ellington—they have led big bands of their own whenever possible. While they have made many small-group recordings, it is as big-band drummers that they have made their most valuable contributions to jazz.

The number of bebop drummers grew dramatically in the 1950s and early 1960s. Building upon the foundations established by their predecessors, these technically proficient and musically inventive drummers played vital roles in expanding the bebop vocabulary. Billy Higgins (born in 1936) is a highly versatile small-group drummer in this second generation of beboppers. He began his recording career in the late 1950s in Los Angeles. At first it appeared that he would become a charter member of the action-jazz school; some of his first recordings were with Ornette Coleman, who in those early recordings was teetering on the edge between bebop and action jazz. Over the years the two men have reunited for other performances and recordings. Later, Higgins recorded with Archie Shepp, Cecil Taylor, and other representatives of post-bebop jazz. Still, he is primarily a bebop drummer, equally at home both with the loudest and most aggressive "hard bop" groups and with the subtlest of groups. He always plays with a crisp precision and elegance. Two of the greatest joys in jazz are to hear him play with brushes—there is no greater master of brush technique—and to hear him support and complement with uncanny perception pianist Cedar Walton, his frequent associate since the 1960s.

It is difficult to isolate specific phrases idiosyncratic to Higgins, for he usually plays well within the mainstream of the bebop vocabulary. But he is fond of an African-sounding ostinato, which he uses sometimes while trading 4s or 8s. In the notated example, the diamond noteheads represent strokes on the counterhoop of the snare drum.

Al "Tootie" Heath (born in 1935), the youngest member of the gifted Heath family (with older brothers Percy and Jimmy), is an equally versatile and musically sensitive drummer. Like Higgins, he has spent most of his professional life as a freelance player. And like Higgins, he dedicates himself to inspiring the group at hand to play as well as possible, no matter what the style, no matter what the environment.

Mel Lewis (1929–90) was also a versatile drummer, but unlike Higgins, Heath, and many other bebop drummers, his first love was the big band, not the small group. He made invaluable contributions to the Los Angeles-based bands of Stan Kenton, Terry Gibbs, and Gerry Mulligan during the 1950s and early 1960s. Then in 1965 he and trumpeter/arranger Thad Jones began co-leading a rehearsal band that played Monday nights at New York City's Village Vanguard. Soon this superb band of studio musicians began recording and going on limited concert tours. Jones left the band in 1979, but Lewis continued to lead it until shortly before his death from cancer. Lewis lacked Rich's and Bellson's flashy technique and showmanship, but his timing, musicianship, and ability to energize a big band were exemplary.

Of the drummers to establish international reputations during the 1950s and early 1960s, the most dramatic were Philly Joe Jones and Elvin Jones (unrelated). Philly Joe Jones (1923–85) had made only a handful of recordings before he became a charter member of Miles Davis's quintet in 1955. Soon he and his rhythm partners Paul Chambers and Red Garland were considered the finest rhythm section in jazz by many in the profession.

Sometimes Philly Joe whispered; he was another master of the brushes.[37] In medium tempos he could lay down a simple groove that swung compellingly but unobtrusively. Under Red Garland's solos he sometimes played nothing more than the basic ride cymbal pattern, the backbeat "chicks" on high hat, and on beat 4, an uplifting cross-stick rim shot on snare drum. But he could, and often did, roar. The sound of his ride cymbal was big and splashy. In fact, much of the time everything he played was big and splashy. He could be the loudest, most aggressive of drummers. On recordings that aspect of his playing is absent, because engineers, not musicians, set the recording balance. In person, however, he all but obliterated the rest of the band in lively pieces, especially during John Coltrane's solos. His loud playing prompted people to tell Davis that he should get another drummer. But Davis wanted Jones's fiery playing, and employed him for three years, until Jones's heroin-driven unreliability became intolerable.

While much of what Jones played derived from Max Roach, Jones had a distinctive approach to double-time playing, often when a soloist began using large numbers of sixteenth notes instead of eighth notes. At such times his ride cymbal remained synchronized to the string bass in $\frac{4}{4}$, while he played his high hat in $\frac{8}{8}$:[38]

Elvin Jones (born in 1927) is only two years younger than Max Roach, but he began recording a decade after Roach did. His first records date from 1953, and include more than three dozen sessions during the 1950s with many prominent bebop players, but his fame rests primarily on his partnership with John Coltrane during 1960–65. Those were years of stretching boundaries for the two men, when Coltrane moved his vocabulary to the edge of action jazz and the two players developed a new soloist-drummer relationship. Jones doubled and quadrupled the amount of accompanimental activity used by Max Roach, and greatly increased the loudness level used by Philly Joe Jones. During Coltrane's lengthy solos Jones churned and boiled relentlessly, both competing with and complementing Coltrane's furious, passionate melodies. The intensity is best heard in concert recordings such as *Impressions,* but some recordings done at Rudy Van Gelder's studio are nearly as intense; among them are *Your Lady* and *A Love Supreme.*[39]

Occasionally Elvin Jones plays solos in free meter, and some of his fills consist of variable-speed rolls on different drums that almost give the effect of free meter. But usually his swinging beat drives forward inexorably and unerringly. It sings out bell-like from his ride cymbal while a thunderous barrage of drum phrases churns all around it non-stop.

During his years with Coltrane, Jones occasionally recorded and concertized with groups that he led, and since leaving Coltrane in 1965 he has led groups most of the time. Often these groups are pianoless trios and quartets, which allow him to explore the textures possible with the tenor sax-bass-drums instrumentation that he, Coltrane, and bassist Jimmy Garrison often explored in the early 1960s.[40]

Tony Williams (born in 1945) astounded the jazz world when he joined the Miles Davis Quintet in 1963, at age 17. From the beginning he brought to Davis a concept of drumming different from those of Davis's previous drummers Philly Joe Jones and Jimmy Cobb. While backing other players he played relatively softly, but used a great array of rhythms that both complemented and reinforced the rhythms of the soloists. Often he played the high hat on every beat, not just on the weak beats. During his solos he often used unmetered rhythms that defy transcription into traditional rhythmic notation. In a detailed study of Williams's early solos, Craig Woodson found that he played rolls "in single-, double-, combination-, buzz-, and flam-stroke stickings," usually on the snare drum, in both measured and unmeasured rhythm, and that the speed of these rolls is anywhere from 3 to over 30 strokes per second. He also used the basic ride-cymbal pattern, but regularly split it among cymbal, snare drum, bass drum, and high hat in a variety of ways. As with rolls, Williams used the ride pattern in both measured and unmeasured time. Finally, Woodson found nearly 200 other rhythmic figures, measured and unmeasured, making up Williams's solo vocabulary (Woodson 1973: I, 91, 121, 130, 136).

The high levels of energy and creativity that Williams brings to his solos he also applies to his accompaniments. *Footprints,* the famous blues performance by the Miles Davis Quintet of 1966 (Columbia 9401), is a fine showcase for his accompanying skills. Ron Carter begins the piece with a $\frac{6}{4}$ ostinato. Williams then enters and reinforces the meter for a short while. Soon, however, he begins to superimpose other complimentary meters on the ostinato. Partway through Davis's solo he plays the ride cymbal patterns in a slow $\frac{2}{2}$; then he moves into double time and forces Carter to modify his $\frac{6}{4}$ ostinato into a syncopated ostinato in $\frac{4}{4}$. While these meter changes unfold, Williams inserts a fascinating assortment of fills. Every member of the quintet plays wonderfully, but the hero of the piece is Williams.

Of all the jazz books waiting to be written—and there are many—none would be more fascinating than a book about the great bassist-drummer teams. The players themselves should write much or all of it. Ray Brown and Ed Thigpen might co-author a chapter, Percy Heath and Connie Kay another, Steve Gilmore and Bill Goodwin another, Ron Carter and Tony Williams another, John Clayton and Jeff Hamilton another, and so on. Perhaps Jimmy Cobb could contribute a chapter on himself and Paul Chambers, Louis Hayes on himself and Sam Jones, Elvin Jones on himself and Jimmy Garrison. This book should go far beyond the level of "Oh, he is (was) a beautiful cat! We had some great times together!" Perhaps it could *start* there, but thousands of those great times involved playing *Our Delight,* or *Bluesology,* or *Footprints,* and having the music soar. What did those team-mates *do* to make the music soar, to give the soloists those effortless rides to great heights of musical inspiration? How did they communicate, cooperate, and create?

An intriguing series of recordings offers us the opportunity to hear the bassist-drummer team in isolation. In the 1970s saxophonist Jamey Aebersold began producing play-along record albums that contain rhythm-section accompaniments for pieces by Parker, Davis, Rollins, and others. The players include such major jazz figures as pianists Kenny Barron, Ronnie Mathews, and Harold Mabern, bassists Sam Jones, Ron Carter, and Rufus Reid, and drummers Louis Hayes, Adam Nussbaum, and Billy Hart. In these recordings, the bass comes through one channel, the piano through the other channel, and the drums are on both channels. Besides offering students superb accompaniments with which to practice improvisation, they provide a fascinating way to study rhythm sections. A case in point is Volume 6 in the series, *All "Bird,"* devoted to Charlie Parker's music.[41] The bassist is Ron Carter, the drummer is Ben Riley. Most of the time these two great players seem perfectly synchronized. But in much of *Yardbird Suite* Carter plays one or two beats in each measure incrementally behind Riley's cymbal strokes. Conversely, at one point in *Confirmation* Riley lags slightly behind the beat while Carter maintains the tempo. Are these subtle discrepancies a result of bringing together two players who seldom play together? Are they the result of an artificial playing situation? (They were accompanying Aebersold, but he is not on the recording, so they must have listened to him on headsets as he played in an isolation booth.) Or are these subtle discrepancies common in jazz? These are difficult questions to answer, because there are so few recordings that allow us to hear bass and drums as a completely isolated pair.

CHAPTER NINE

Other Instrumentalists

In the previous six chapters I discussed players of the most common bebop instruments. While the main features of the bebop language appear in their music, there are great musicians who play other instruments. They have made vital contributions to the music, both through their notes and through the varied instrumental timbres with which they enliven some of bebop's finest ensembles. For most instruments, the number of first-rank players is relatively small, so I have grouped them into this single chapter. The result, though a potpourri, is hardly more varied than, for example, the remarkable array of dissimilar musicians in Chapter 7, who are grouped together only because they happened to be pianists.

Woodwind Players

Baritone saxophone soloists form a small group compared with their alto- and tenor-playing colleagues. The "bari" is an almost indispensable component in big bands, but few of those section players are outstanding soloists. In the swing era Harry Carney in the Ellington band and Jack Washington in the Basie band were the best known; in early bebop the main players were Serge Chaloff (1923–57), Leo Parker (1925–62), Cecil Payne (born in 1922), and Gerry Mulligan (born in 1927). In selecting role models the beboppers largely ignored the early baritone players, and turned instead to Lester Young, and especially to Charlie Parker.

The most famous bebop baritone saxophonist is Gerry Mulligan, who is also an excellent composer and arranger. He participated in Miles Davis's nonet recordings in 1949–50 (see pp. 23–24). His fame grew rapidly after he formed his first quartet in 1952 (see pp. 210–11), and he gained additional renown with his Concert Jazz Band in the early 1960s.

Mulligan's improvised melodies are simple, straightforward, and unerringly swinging; they are in the spirit, if not the style, of Lester Young. In playing his infectiously swinging melodies he often avoids the lower third of his instrument's range, preferring instead to play notes well within the range of the tenor saxophone. And in contrast to most baritone saxophonists, he always has favored a light, airy tone quality. Over the years he has refined the tone quality into one of great beauty.[1]

After the first small crop of bebop baritone saxophonists entered jazz, a few others earned recognition; chief among them were Lars Gullin (1928–76), Pepper Adams (1930–86), and multiple reed player Nick Brignola (born in 1936). Like Mulligan, Gullin had a light, airy sound, and in the 1950s he had a melodic style influenced by Lee Konitz. Brignola and Adams both preferred a heavy, gruff tone quality and an aggressive, active melodic style. Adams, the best known of these three players, co-led a quintet with Donald Byrd during 1958–64. He spent many years in big bands, especially the Thad Jones/Mel Lewis band. His skill and cleverness in developing a single motive over a long span of time was unsurpassed by anyone in jazz.

In early jazz styles, the clarinet was an essential instrument, but during the 1940s and 1950s, beboppers seemed to reject it. While clarinetists Buddy DeFranco, Jimmy Giuffre, John LaPorta, and Tony Scott played bebop when they could, they had to double on saxophones or play older jazz styles much of the time. Several saxophonists doubled on clarinet occasionally—Buddy Collette, Art Pepper, Phil Woods, and more recently Paquito D'Rivera—though they did not commit themselves to a career on the instrument.

In the 1980s Eddie Daniels (born in 1941) became the foremost bebop clarinetist, a position he continues to hold in the early 1990s. He established his reputation as a saxophonist in the 1960s, playing in the Thad Jones/Mel Lewis band, and as a versatile woodwind player on many New York studio sessions. At the same time he earned a graduate degree in clarinet at the Juilliard School of Music, and within a few years was devoting his career exclusively to has favorite instrument. In recordings such as *To Bird with Love*,[2] he combines an astonishing technical command of the instrument, a beautiful, warm tone quality over the entire range of the instrument, a great sense of swing, and a rich melodic imagination. He has proven repeatedly and conclusively that bebop can fit the clarinet; the only barrier to a flood of bebop clarinetists appearing may be the challenge of meeting Daniels's awesome standards.[3]

The flute—all but unknown in jazz until the 1950s—has never been a major jazz instrument. During the past 40 years, most flute solos have been played by musicians wearing saxophone neck straps—Buddy Collette, Eddie Daniels, Rahsaan Roland Kirk, Moe Koffman, Yusef Lateef, Charles Lloyd, James Moody, Sam Most, Jerome Richardson, Bud Shank, Sahib Shihab, Lew Tabackin, Frank Wess, and others. Herbie Mann, Jeremy Steig, James Newton, Steve Kujala, and Hubert Laws are among the few full-time flutists in jazz. Typically the flute is associated with quiet and lighthearted music—it was a common instrument in some "cool" groups of the 1950s. But Kirk, Tabackin, and others have created aggressive, emotional effects by humming or singing while playing, and Steig regularly expands his palette of timbres with percussive key clicks and high-pitched air sounds.

The chromatic harmonica would have no place in jazz were it not for Toots Thielemans (born in 1922). In the early 1950s, during his tenure as the guitarist in the George Shearing Quintet, he recorded a few pieces on harmonica. His success with these early recordings led to a lifetime of international concert appearances and recording sessions. His readily identifiable sound, with its expressive pitch bends reminiscent of Miles Davis, graces recordings by vocalists, big bands, and small groups in both the jazz and popular music worlds. The jazz world knows him as a gifted bebopper; preschoolers past and present know him from the closing theme of *Sesame Street.*

Brass Players

J. J. Johnson (born in 1924) and Kai Winding (1922–83) were the first influential bebop trombonists. Johnson, while a member of Benny Carter's big band, recorded some lengthy solos in the first official Jazz at the Philharmonic concert in Los Angeles in 1944. At the time his exuberant but immature swing-oriented style contained only small traces of bebop. The following year he joined the Basie band, and also began leading small-group recording sessions. By then he had learned many of Gillespie's and Parker's melodic formulas. His solo work on *Jay Bird, Coppin' the Bop, Jay Jay,* and *Mad BeBop* (typical early bebop contrafacts of *I Got Rhythm, Honeysuckle Rose,* and *Just You, Just Me*) provided models for other trombonists to emulate while moving into the new idiom. His playing in these pieces was stylistically consistent and fluent, though there was a roughness to some of his attacks and transitions from note to note. He smoothed out these attacks and transitions very quickly, though, and generally improved his technical facility to the point that he sounded as if he were manipulating valves rather than a slide (for example, *Bone-o-logy* of 1947). He further enhanced this unheard-of smoothness and dexterity by using a mute frequently.[4]

Johnson developed a style characterized by singable melodies played with precise attacks on individually tongued notes, a light vibrato used as an occasional ornament, and a rhythmic security that makes his melodies swing, with or without a rhythm section. Some of his vocabulary is directly out of Parker's, such as Figures 1A, 2A, 3, and 4, and the descending scalar organization (all discussed on pp. 31–36):

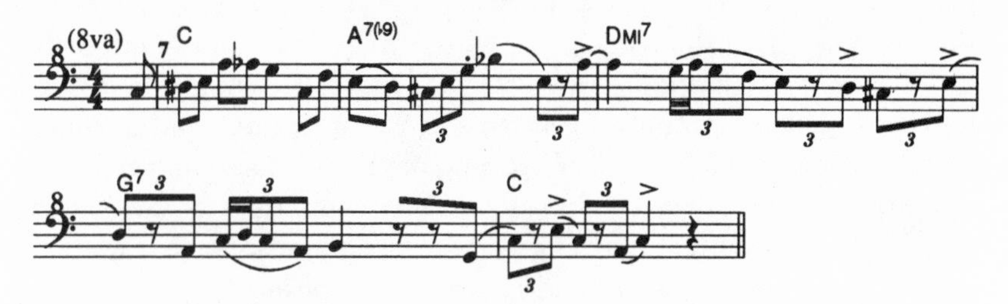

But he often uses two- and three-note bits of melody instead of Parker's long lines in eighth notes. And much of his vocabulary includes elements idiomatic to the trombone, such as tongued repeated notes—

—or fall-offs and smears produced with the slide.[5]

While Johnson was establishing his credentials, Kai Winding was building a reputation in the big bands of Benny Goodman (1945–46) and Stan Kenton (1946–47). He also assimilated the basics of Gillespie's and Parker's vocabularies, but slightly later than Johnson. Unlike Johnson, Winding retained a roughness to his tone and attack, and when listening to his early solos one is always aware that he is playing a slide, not a valve, trombone.

During 1954–56 these two bebop trombonists co-led a successful quintet, Jay and Kai. By then their styles owed less to those of Parker and Gillespie, for they had developed a trombonists' approach to playing bebop. After 1956, they enjoyed occasional reunions over the years, but otherwise went separate ways. Winding spent most of his career as a player, and remained an energetic bebop stylist for the rest of his life. In the late 1950s Johnson led a series of superb quartets and quintets that included at various times cornetist Nat Adderley, trumpeter Freddie Hubbard, pianists Tommy Flanagan and Cedar Walton, and drummer Tootie Heath. In addition, however, he developed a second career as a composer. He had developed an interest in composing as early as the 1940s, and had written or arranged most of the music for Jay and Kai. By the late 1960s he was focusing more on composing, less on playing. But twenty years later, after notable success as a Hollywood film and television composer, he returned with new vigor to his first love. His 1988 albums *Quintergy* and *Standards,* recorded in concert at the Village Vanguard, show conclusively that his technical facility and solo imagination remain undimmed.[6] Now a septuagenarian, he remains the most admired and influential trombonist in bebop.

Winding was the first of several bebop trombone soloists who played in the Kenton band during the late 1940s and 1950s; others included Milt Bernhart, Eddie Bert, Frank Rosolino, Carl Fontana, Herbie Harper, and bass trombonist George Roberts. Of this group Rosolino and Fontana, both inventive and facile players, were the most prominent. Other important trombone soloists who emerged in the 1950s included Jimmy Cleveland, Benny Powell, Curtis Fuller (with The Jazztet, then with Art Blakey), Slide Hampton (both a composer and a player), Jimmy Knepper (with Mingus, then with Gil Evans), and David Baker (with George Russell in the early 1960s; a professor at Indiana University since 1966, and a famous jazz educator). In the 1970s Bill Watrous established a solid reputation as a virtuoso; he has a thinner tone quality than his predecessors, but possesses a phenomenal technique that makes his playing in the highest range of the instrument sound easy.

There also are some fine players who specialize in the larger valve instruments. The best of the valve trombonists include Bob Brookmeyer, Bob Enevoldsen, and Rob McConnell. Cy Touff has been playing the bass trumpet for over 40 years; he still has it all to himself. The French horn is more popular; lately the bebop pioneers John Graas, Julius Watkins, and Willie Ruff have been joined by Bobby Routch and Richard Todd. And in the world of baritone horns, euphoniums, and tubas, Don Butterfield, Howard Johnson, Rich Matteson, and Harvey Phillips solo with an agility and command of the idiom equal to that of the best string bassists.

Vibraphonists

Milt Jackson (born in 1923) surely heads anyone's list of bebop vibraphonists. He began performing with Dizzy Gillespie's groups in the mid-1940s, and has been one of bebop's greatest soloists for over 40 years. Besides being a charter member of the Modern Jazz Quartet (see pp. 211–14), he has made a vast amount of fine recordings as a leader and as a member of all-star ensembles.

From the beginning, Jackson looked to his colleagues in bebop for inspiration, not to Lionel Hampton or Red Norvo, the most prominent swing-style vibraphonists. His vocabulary includes many of the melodic figures shown in Chapter 3, figures shared by saxophonists, pianists, and others. He also uses some figures that are less widespread. Especially ear-catching are phrases incorporating short notes that are nearly grace notes:

(Chick Corea favors a similar ornamentation—see pp. 165–66.) Perhaps his figure that is most idiomatic to the instrument is a series of repeated notes, sometimes in triplet eighths, sometimes in sixteenths:[7]

Compared with Parker, Gillespie, Powell, and other early bebop leaders, Jackson uses more triplet eighth notes in his improvisations. He also makes greater use of rhythmic and dynamic contrasts, especially in ballads, where his rhythmic flexibility and general expressivity is marvelous to hear. The slow speed of his instrument's oscillator is an important component in his expressive style. He runs it at about 3.3 revolutions per second, instead of the 10 rpm used by Hampton. The result is that his long notes have a beautiful, subtle motion instead of the nervous shimmy that originally was the norm on the vibraphone. He often exploits that beautiful sound by ending a piece with a slow arpeggio of a simple major triad, letting the notes ring for several seconds.

Many bebop vibraphonists looked to Jackson for inspiration—Terry Gibbs, Cal Tjader, Victor Feldman, Dave Pike, and others. Bobby Hutcherson (born in 1941) also emulated Jackson at first. But he soon developed a distinctive style, one that utilizes a rapid-fire melodic style that sometimes seems the equivalent of John Coltrane's "sheets of sound." His rapid runs played without the sustaining pedal have a dry, clear sound that suggests the xylophone or marimba; in fact, he is the only prominent bebop vibraphonist to double regularly on the marimba.

The bebop vibraphonist whose style differs most from Jackson's is Gary Burton (born in 1943). One obvious difference between the two players is that Burton uses no oscillation; his compact, fold-away instrument (efficiently engineered for road trips) does not even have a motor—a vibrato-less vibraphone. A second difference is his usage of four mallets (most players use two), which he uses to create polyphonic textures nearly as dense as those created by a solo pianist. (Not surprisingly, his principal role model was not a vibraphonist, but pianist Bill Evans.) In the mid-1960s, Burton's four-mallet virtuosity enabled him to be the chord player in Stan Getz's quartet, and later allowed him to play solo vibraphone for entire programs.

Electric Organists

Organists make up a small but important part of the chord players of bebop. Almost without exception they play the Hammond electric organ, invented in the 1930s as an inexpensive alternative to the pipe organ. Players create timbres on the Hammond by mixing nearly pure tones pitched at nine different frequencies (most correspond to the natural overtone series). These tones are controlled by drawbars, each of which adjusts to eight different loudness levels. Although the instrument sounds little like a pipe organ, it quickly found a niche in the church music world, and became an essential ingredient in black gospel music. It also gained acceptance in the jazz world, when Count Basie, Fats Waller, and others began to use it occasionally. In the early 1950s, Wild Bill Davis and Milt Buckner were among the first to devote their careers to this instrument, and they established the organ trio instrumentation of organ, guitar, and drums.

During the early 1950s, pianist Jimmy Smith (born in 1925) began playing the organ. In the mid-1950s he made his first records with Don Gardner's trio, using some playing techniques (that is, locked-hands chords, bass line in the pedals) and the same drawbar setting (all nine drawbars at maximum loudness) that Buckner and Davis used. But by early 1956, when he made his first Blue Note album as leader,[8] he had shifted most of his bass lines to his left hand on the lower manual, had developed a single-note melodic style based on long strings of even eighth notes played staccato, and began developing some more transparent alternatives to the muddy drawbar settings he had borrowed from his predecessors. Soon he acquired the new B3 model of Hammond organ, which contains some percussion effects for the upper manual. The extra bite these effects provide has been an integral part of his style ever since. By the end of the decade, his mature style fully in place, he was a formidable master of the instrument, able to synchronize perfectly his left hand with a drummer while improvising intricate melodies:[9]

Smith's impact was large; nearly every other organist to enter the jazz world owes much of his/her style to Smith. First, Brother Jack McDuff, Shirley Scott, and Don Patterson established themselves; then, in the 1960s, came Richard "Groove" Holmes, Jimmy McGriff, Mike Carr, Art Hillery, Charles Kynard, and Big John Patton. Even Larry Young (1940–78), whose melodic lines are closer to those of McCoy Tyner, took his articulation habits and his main timbres from Smith. Today, as in the late 1950s, Jimmy Smith's style is the standard against which all other organ styles are measured.

Guitarists

Many bebop rhythm sections have a pianist but no guitarist; other rhythm sections have a guitarist but no pianist. Since the primary job of both instrumentalists is to comp, and since comping procedures are largely the same on both instruments, there is usually no need to have them both. Indeed, a second comper can become an unwelcome redundancy, as the two chord players select conflicting chord extensions and substitutions, and bump into one another's rhythmic punctuations. In the studio, skillful recording and mixing can minimize most of the damage done by unyielding, competitive compers, but in performance the players must resolve the conflicts themselves. To do so they must alter their standard approach to comping. They must listen carefully, not only to the soloist and drummer, but to each other, and make many adjustments. Sensitive players sometimes take turns comping and resting ("laying out"). One player may play quiet, sustained chordal backgrounds while the other comps more vigorously, or one may play a background riff while the other comps. On some tunes the old swing technique of four-beat guitar strumming (playing "rhythm guitar") solves the problem, for it allows the pianist to comp freely. The wonderfully compatible guitarist-pianist teams of Herb Ellis and Oscar Peterson, Joe Pass and Oscar Peterson, and Ed Bickert and Jimmy Dale (in Rob McConnell's big band) have used all of these solutions repeatedly, and have set the standards for this segment of the art.

In the 1950s many guitarists found a comfortable niche in organ trios (organ-guitar-drums) and quartets (tenor saxophone-organ-guitar-drums, typically). In these groups the organist, who is usually the leader, serves as both the main soloist and the bassist. During organ solos the guitarist comps, and during guitar solos the organist comps, so conflicts are minimal. On occasion some bebop guitarists have chosen to play unaccompanied. This option forces them to supplement the normal single-note melodic approach with occasional block-chord punctuations, longer block-chord passages, two-voice counterpoint, or some combination of these textures. The most acclaimed master of solo guitar performance is Joe Pass, although Barney Kessel, Jimmy Raney, Lenny Breau, and others have also made excellent unaccompanied musical statements. Jazz, however, is primarily ensemble music, and guitarists generally function most effectively as part of a team. Because of the chordal nature of the instrument, many guitarists feel comfortable in a trio setting with a bassist and a drummer. The texture of these trios is generally lighter, and the dynamic range smaller, than those of piano trios, but the music is often no less compelling.

The guitarists who were active during the early stages of bebop looked to Charlie Christian (1916–42) as their principal role model, just as the early bebop tenor saxophonists emulated Lester Young and early bebop bassists emulated Jimmy Blanton. Christian, the first great player of the new hollow-body electric guitar, was not a bebop player any more than Blanton was; both men succumbed to tuberculosis at a tragically young age before bebop blossomed. He was part of the milieu at Minton's and Monroe's during the critical transition years, however, and had he lived he surely would have adapted to the new idiom. His influence on early bebop guitarists stemmed less from his after-hours associations—only a handful of bebop guitarists heard him in those Harlem jam sessions—than from his position as the most admired guitarist in jazz. His consistently inventive solos on recordings with Benny Goodman's sextet were famous.

The first guitarists associated with bebop included Chuck Wayne, Arv Garrison, Remo Palmieri, Billy Bauer, Mary Osborne, Bill DeArango, John Collins, and Barney Kessel. At first their main functions were to play rhythm guitar and written-out unison lines with other lead instruments. Their brief solos usually showed a strong influence of Charlie Christian, though some players, especially Garrison and DeArango, found ways to deviate from Christian's model.

Billy Bauer (born in 1915) was another who departed from the straightforward harmonic approach of Christian. As an associate of pianist Lennie Tristano he adopted the more dissonant but rhythmically blander idiom of the Tristano school. (In the 1950s, after his association with Tristano had ended, he sometimes reverted to the Christian style, however.) He was one of the few guitarists in the 1940s who comped, even when a pianist was also comping. Unfortunately, during that time neither Tristano nor Bauer was a sympathetic accompanist; they competed, not comped, and produced a cluttered sound.

During his teenage years Barney Kessel (born in 1923) played informally with Christian in their native state of Oklahoma. At first he based his style on Christian's, though his tone quality, characterized by a noticeable pick noise, deviated from his role model. But by 1947 he had picked up many bebop licks and articulation habits—his role models seemed to be Parker and other wind players—and had become a supportive and propulsive comper. For the next several years his style was a synthesis of long bebop lines and Christian-like licks, such as the cadential figure borrowed from Christian:

The noisy articulation has remained a characteristic of his sound, and for him precise coordination of fingering and picking in rapid-note phrases seems less important than the overall rhythmic gesture. Kessel's early career alternated between periods of prominence in the jazz world (with Artie Shaw in 1945, with the Oscar Peterson Trio in 1952–53) and years of financial success as an anonymous recording session player. Since the late 1960s his primary activities have again been outside the studios.

Although Kessel began to adopt the bebop idiom in the late 1940s, he was not well known as a bebop guitarist until the 1950s. In fact, the guitar's place in bebop was not fully defined until that time, when many fine guitarists joined their predecessors. Among the best were Ed Bickert, Kenny Burrell, Herb Ellis, Grant Green, Jim Hall, Mundell Lowe, Wes Montgomery, Jimmy Raney, Howard Roberts, Toots Thielemans (known primarily as the world's most prominent jazz harmonica player), and Al Viola. In the 1960s Barry Galbraith emerged occasionally from long years of studio work to make some important solo statements, and Joe Pass and George Benson began their ascents to the top echelon of bebop guitarists.

Among this fine group of players none developed a more distinctive style than Wes Montgomery (1925–68). His sound partly stemmed from the simplest reason; instead of using a plastic pick, he plucked the strings with the fleshy part of his thumb. The immediate result was a softer, gentler attack than that achieved by his colleagues. The other immediate style identifier was his frequent habit of playing melodies in parallel octaves (and sometimes parallel fifteenths) with a seemingly effortless facility:[10]

In the final analysis, however, we value Montgomery not for the gentle attack of his "golden thumb" or for his octave acrobatics, but for the great melodic and rhythmic inventiveness that he displayed in his best recordings—those done with small bebop groups. Writers in the 1960s hailed him as the most inventive guitarist since Charlie Christian, and that view is hard to dispute.[11]

Joe Pass (1929–94) might have entered the international jazz scene as early as the early 1950s. Unfortunately, drug usage delayed that entry for a decade, until the album *Sounds of Synanon* appeared (named after the drug rehabilitation program in Santa Monica, California, that helped him regain control of his life).[12] Subsequent albums—with Richard "Groove" Holmes, Johnny Griffin, Les McCann, Gerald Wilson, and several of his own—established his importance as a top bebop guitarist. In the 1970s his star rose ever higher, because of important albums with Carmen McRae, Duke Ellington, Oscar Peterson, Ella Fitzgerald, and Herb Ellis, and perhaps most significant, because of a series of solo albums, beginning with *Virtuoso.*[13] From 1973 he was a Pablo recording artist, joining Peterson, Fitzgerald, and the others under Norman Granz's sponsorship.

Pass's remarkable success was well deserved, for he was a supremely gifted musician. His solo style was a guitarist's parallel to that of Clifford Brown—not because of melodic similarities, but because, like Brown, Pass was consistently inventive, seldom flaunted his considerable technical skills, had a beautiful tone quality and clean articulation, and swung unerringly. In an ensemble he was the most supportive yet unobtrusive of compers. As an accompanist of singers he simply had no superior, and few if any peers. And his gently swinging unaccompanied playing was a marvel of good taste and inventiveness.

George Benson (born in 1943) began his recording career as a jazz guitarist in the early 1960s, in Jack McDuff's organ quartet.[14] He soon established himself as a creative bebop guitarist with an unusually fast and precise technique. In 1966 he began singing on some of his albums, and by the 1970s his expressive tenor voice was as well known to his jazz club and record audiences as was his brilliant guitar style. On some pieces he combined both talents, by singing and playing his improvised lines in unison. Within ten years his audience increased enormously. In 1976, his album *Breezin'* sold over 1,000,000 copies—a spectacular figure for a jazz record. The hit song from that album was an edited version of *This Masquerade,* essentially a contrafact of Matt Dennis's *Angel Eyes.* Benson's recording, the full-length version of which features his tenor voice singing the song, his wordless guitar-voice unison, and some guitar solo lines, won the Grammy award for Record of the Year, given by the National Academy of Recording Arts and Sciences in early 1977. Other hit records followed, including *On Broadway,* which also features his voice-guitar unison lines.[15] Thus Benson discovered, as Nat Cole had done years earlier, that singing is much more lucrative than playing. Nevertheless, he has maintained his guitar skills, as evidenced occasionally on his albums and on those of others. He remains among the world's finest guitarists.

CHAPTER TEN

Ensembles

An assemblage of great jazz musicians will not always make great jazz. Different players, though they may all play within the bebop idiom, develop different concepts of rhythm, harmonic vocabulary, texture, and dynamics. Sometimes those concepts are too different to be compatible. If they are potentially compatible, the players still may need some time together—weeks, months, even years—to reconcile their differences and learn to function in a cohesive and mutually supportive way. A new group has many questions to answer for each piece: What is the best tempo? Should there be a written arrangement or will a head arrangement suffice? In either case, what is the proper interpretation—dynamics, articulations, phrasing—of the arrangement? What kind of accompaniments are best for the different soloists? How is the ending to be played, especially if it is out of tempo?

The players must answer these and other similar questions, verbally or non-verbally, or in both ways, before they can make great ensemble music.[1] Once they answer the questions and their music-making moves to a higher level of spontaneous creativity, their musical interaction may become among the most intense and rewarding of human relationships. The emotions felt by the musicians in such close communication with one another can be as significant as the emotions felt by lovers for each other or by parents for their child. Those of us on the outside of this intense musical communication may not share in all those emotions, but we surely can experience the joy of hearing the fine music that results from this communication. In this chapter I focus on some of those great, communicating ensembles and their music. The ensembles range in size from trios to big bands.

Bud Powell's most frequent ensemble was a trio. Because he worked with several different bassists and drummers and because his mental health was unstable, he did not always produce the best in ensemble music. But on those occasions when he and his colleagues were functioning at a peak levels, superior music resulted. One classic example is *Un poco loco,* played by Powell, Max Roach, and Curly Russell.[2] Blue Note records issued three takes of this piece, which give us a fascinating glimpse into the ensemble creative process. The structure of the performance is as follows:

introduction (8 measures),
first theme in *aaba* form (32 measures),
codetta (8 measures),
transition (8 or 4 measures),
second theme (32 measures),
Powell's improvisation on a tonic pedal (open-ended),
Roach's solo (open-ended),
first theme,
codetta.

As the title implies, this piece has a Latin-American flavor. In take 2, the first issued take, Roach uses a common Latin pattern:

In this take, Powell plays some ideas that he probably had been using as he practiced the piece, for they reappear in the subsequent takes. Yet the solo does not gel, and he aborts the piece before the return to the first theme. In take 3, Roach tries a radically different pattern:

This polyrhythmic pattern transforms the piece from a run-of-the-mill Latin-jazz piece into something unique.[3] The take is complete, ending with a half-chorus theme. Though it was a successful performance, the players wisely decided to play it again; take 4, the best of the takes, was the result. Powell's 100-measure solo over the tonic pedal works well, as does Roach's 34-measure solo. The result is a jazz landmark, marred only by the undermiking of bassist Russell. Because of the undermiking, the performance sounds more like a duet than a trio, but is a wonderful ensemble piece nonetheless.

Oscar Peterson established his worldwide reputation in the 1950s, playing primarily in trios that he led. His trios, unlike those of Bud Powell, stayed together for long periods of time, enabling the players to develop a high degree of musical rapport. His early trios adhered to the instrumentation established by his early idol, Nat Cole: piano, guitar, and bass. For fifteen years his bassist was the powerfully swinging Ray Brown; his guitarists were Barney Kessel, Irving Ashby, and—from 1953 to 1958—Herb Ellis.

The Peterson-Ellis-Brown trio marked a turning point in Peterson's musical evolution. The earlier trios were Peterson and two accompanists, but this group was truly an ensemble. They concertized extensively and rehearsed endlessly to develop and perfect their head arrangements (they never worked from written arrangements), and achieved remarkable results. *Swinging on a Star,* from a concert at Carnegie Hall, is an early example of their great ensemble playing. This Jimmy Van Heusen popular song of the 1940s is a bit unusual. The lyrics follow the usual verse-refrain format, but since the refrain is only 8 measures long, the 12-measure verse is always sung (or played), resulting in a 20-measure chorus. The group plays a complicated and well-rehearsed introduction. Ellis then plays the verse, and Peterson the refrain. A transition leads to split choruses by Brown (verses) and Ellis (refrains), then several full choruses by Peterson that build in intensity. The closing chorus and coda are complex and well worked-out. While Peterson is clearly the star—he has five choruses, while Brown and Ellis split three—the complexity of the arrangement and its flawless execution result from three gifted artists working together intently.

In 1958, Ellis, tiring of life on the road, left the trio. Peterson decided to replace him with a drummer instead of with another guitarist. Gene Gammage held the job briefly; then in the spring of 1959 Ed Thigpen joined the trio, and completed another classic ensemble. For the next six years this superlative piano trio concertized and recorded extensively. Again, this trio played as an ensemble, with well-devised arrangements and shared solo duties. *Tangerine* is but one of many fine examples of their art.

When Brown and Thigpen left in the mid-1960s, Sam Jones and Bobby Durham joined Peterson, keeping the ensemble tradition going at the same high standard until 1970. Since then there have been other Peterson trios and quartets, some involving reunions with former colleagues, some with different world-class bassists, guitarists, and drummers. Always the musical standards and sense of ensemble are high, for Peterson dislikes unprepared, jam-session performances with his groups.

Peterson's recorded repertory is enormous. It includes entire albums devoted to individual songwriters (Ellington, Porter, Gershwin, Kern, and others), albums devoted to theatrical productions (*My Fair Lady, Porgy and Bess, West Side Story*), and hundreds of other pieces drawn from the swing and bebop repertories, from the popular song world, and from his own compositions. Like Art Tatum, with whom he is compared often, he has no trouble learning new pieces and developing something interesting to do with them.[4]

Bill Evans, like Powell and Peterson, preferred the trio format. His relatively subdued style demanded a whole different set of group dynamics from those found in Peterson's trios, but the resulting music was just as great. The first important trio included bassist Scott LaFaro and drummer Paul Motian in 1959–61 (see pp. 160–61, 176–77). The relationship between Evans and LaFaro often went beyond that of soloist and accompanist; sometimes they functioned as equal melodic partners engaged in fascinating dialogues. In their early album *Portrait in Jazz* they set high standards of group interplay with *Autumn Leaves, Peri's Scope, What Is This Thing Called Love?, Someday My Prince Will Come, Blue in Green,* and others. It is a landmark album in the history of piano trios.

This trio's best recordings are the final ones, made in concert at the Village Vanguard a few days before LaFaro's death. The joys are many here, including the impeccable and compelling sense of swing even when no player is explicitly articulating the beats, the ESP-like communication that seemed to guide Evans and LaFaro in the out-of-meter ending of *Waltz for Debbie,* take 2, the musical dialogue near the beginning of *My Romance,* take 1, and, especially, the amazing improvised counterpoint of *Solar.*

Evans begins *Solar* alone, playing a skeletal and rhythmically ambiguous version of Miles Davis's famous 12-measure non-blues theme. LaFaro enters for the last four measures, playing a bit of the theme and then extending its descending scalar motive downward while Evans moves toward his first improvised chorus. In this second chorus, LaFaro develops the initial four-note motive of the theme, while Evans goes in a different melodic direction. In the next two choruses LaFaro plays rhythmic alterations of the theme as Evans continues his improvisation. After the fifth chorus he abandons the theme almost entirely—although fragments pop up unexpectedly here and there—but continues his non-walking bass lines. Evans ingeniously spins out some rhythmic motives, and all three players dance lightly around the beat and the harmonic structure. Because they obviously all hear the beat in the same way, they swing, though in a much different way from the way Oscar Peterson's trios swing. For most of Evans's first solo he plays a single melody in parallel octaves, so the harmonies are implied, not explicit. The listener must concentrate to keep from getting lost, for the players scatter the structural and metrical markers unpredictably during these first 16 choruses. The next 15 choruses—LaFaro's solo—demand even closer scrutiny. Partway through the solo Evans drops out, and LaFaro and Motian do little to articulate the chorus endings explicitly. Eight choruses of Evans and Motian trading choruses follow, always with LaFaro pausing, hopping, running—everything but walking. Then come two cryptic theme choruses, a short coda, and the piece ends. It is a complex piece that took great skill and teamwork to create, and demands informed and repeated listening on our part.

Evans had other fine trios, including one with bassist Chuck Israels and drummer Larry Bunker, and one with bassist Eddie Gomez and drummer Eliot Zigmund.[5] But the finest was with Gomez and drummer Marty Morell. Gomez, as consummate a bassist as Evans was a pianist, played with Evans during the late 1960s and most of the 1970s. Their rapport equaled and at times exceeded anything Evans and LaFaro realized. During the early 1970s, when the drummer was Morell, the ensemble playing in *My Romance, Turn Out the Stars,* and other pieces was outstanding.

Another type of ensemble emerged from Evans's trio album, *Conversations with Myself* (1963), his duet album, *Further Conversations with Myself* (1967), and his duet-and-trio album, *New Conversations* (1978), all done by means of overdubbing.[6] In the second album, aided by the stereo separation of the recording, Evans Left solos while Evans Right comps, and vice versa. The two Evanses are complementary, not competitive; for once the second piano is not redundant. In the third album, he again plays some effective duets and trios with himself, using two grand pianos or two grands and a Rhodes keyboard. Even in the trios he is complementary, not confusing or competitive. His ability to replicate crescendos and rubato tempos during overdubbing, especially on *Song for Helen,* is uncanny. This is another landmark album, beautifully played and beautifully recorded.[7]

The typical bebop quartet is a piano trio plus a wind instrument—usually a saxophone. During the 1950s and 1960s one of the world's most popular bebop quartets, the Dave Brubeck Quartet, had that instrumentation. From 1951 until 1967, and intermittently during the 1970s, Brubeck's saxophonist was Paul Desmond (see pp. 68–70). The bassist and drummer changed frequently at first until Gene Wright and Joe Morello settled into place in the late 1950s.

Brubeck (born in 1920) is a much different pianist from either

Peterson or Evans. In the early years of the Quartet his playing often exhibited a Jekyll-and-Hyde duality; during his solos he tended toward bombast rather than grace, and pseudo-Baroque-style noodling rather than straightforward swing. Yet he was a sympathetic and supportive comper. Time and again the musical high points of the Quartet's recordings are Desmond's elegant melodic lines garnished with Brubeck's gentle harmonic commentary. Desmond's solo in the simple blues *Balcony Rock* illustrates this teamwork beautifully (see also Chapter 4).[8] Near the end of the performance, after Desmond and Brubeck have played multichorus solos, they reveal another kind of teamwork. For one chorus they engage in imitative collective improvisation. It is another effective moment, but it is over almost before the concert audience realizes what happened.

In 1952, baritone saxophonist Gerry Mulligan formed a piano-less, guitar-less, organ-less quartet: baritone saxophone, trumpet, bass, and drums. This novel group quickly became popular in the jazz world, especially in California. The keys to the group's success were Mulligan, with his beautiful, lyrical tone and soft, relaxed swing, and his first trumpeter, Chet Baker, a lyrical and uncommonly subdued trumpeter. They blended well and developed some effective, emotionally low-keyed arrangements that found instant acceptance. But the blend was even better with Baker's replacement, valve trombonist Bob Brookmeyer (born in 1929). In their first recording together—a concert performance in Paris—they display a wonderful sense of dynamic and stylistic balance, especially in Mulligan's composition *Walkin' Shoes.*[9] Mulligan's engaging *aaba'* theme begins with a simple, swing-style phrase, but some of the harmonic structure, especially in the bridge, has an intriguing complexity that goes beyond the norms of swing style. His arrangement is simple but effective, and gracefully executed. Unlike the typical unison theme statement of earlier bebop, it contains two-part writing here and there to add variety. (Such writing became an important part of small group performances after the early 1950s. Art Blakey's Jazz Messengers, Shelly Manne and His Men, the Cannonball Adderley Quintet and Sextet, and especially Horace Silver's Quintet were among the best of the small ensembles of the 1950s and 1960s who played arrangements, not just themes and solo choruses.) This arrangement ends with the ultimate two-measure break—complete silence. Mulligan's subsequent one-chorus solo is a model of subtle swing. His harmonically explicit melodic lines, coupled with Red Mitchell's fine walking bass lines and a simple countermelody by Brookmeyer, make us forget that the group has no chord player. In the next chorus Brookmeyer and Mulligan trade roles as soloist and accompanist. Then at the start of the fourth chorus Mulligan engages Brookmeyer in a non-imitative but rhythmically complimentary dialogue for half a chorus. They return to the theme at the bridge and press on to the coda, during which Mitchell plays some effective two-measure breaks. The out-of-tempo ending is obviously well rehearsed. The performance is emotionally low-key but thoroughly convincing, and the Parisian audience clearly enjoyed it.

The Modern Jazz Quartet, officially organized in the early 1950s, occupies a unique position in jazz history. Its pre-history began in 1946, when vibraphonist Milt Jackson, pianist John Lewis, bassist Ray Brown, and drummer Kenny Clarke were colleagues in Dizzy Gillespie's big band. These four men played as a quartet during the big band's performances, to give the brass players rest breaks after playing the more difficult arrangements. Five years later these four men recorded as the Milt Jackson Quartet. Late in 1952, Jackson, Lewis, Clarke, and bassist Percy Heath made the first recording under the name Modern Jazz Quartet. At first just a recording group, the MJQ soon became a working ensemble, with Lewis functioning as musical director and chief composer/arranger. In 1955 Connie Kay replaced Clarke as the drummer, and to this day remains the newest member of the Quartet. They disbanded temporarily in 1974, but soon resumed a limited annual schedule of concert tours. Now nearing the end of their fourth decade together, Jackson, Lewis, Heath, and Kay have maintained an unchanged personnel longer than any other jazz group.[10]

The MJQ's beautiful ensemble sound is instantly recognizable, largely because of the timbre of Jackson's vibraphone. Jackson, the most emotionally expressive vibes player in jazz, seems able to improvise an endless supply of rhythmically flexible, blues-tinged melodies. The overall style of the Quartet is restrained, guided by Lewis's light touch and conservative melodic and harmonic vocabulary. Lewis likes to play simple countermelodies during Jackson's solos instead of the usual chordal comping, and often spins delicate and beautifully unified melodies during his own solos. Heath and Kay have played together so long they often seem to think as one. Together the four men weave some of the most remarkable polyphonic fabrics in jazz.

The MJQ's repertory has much in common with repertories of groups discussed earlier: popular standard songs, swing and bebop standards, and compositions by Jackson. The blues (many written by Jackson) make up a large and important part of this repertory, and the quartet plays them superbly. The main difference between the quartet's repertory and that of other groups is the large body of compositions that Lewis has written. Some are straightforward themes, used for choruses of improvisations. Others are more elaborate multisectional single pieces; still others are suites of pieces in the tradition, but not the style, of Duke Ellington's larger works.

Django is Lewis's touching elegy for swing guitarist Django Reinhardt. There are two main parts to this composition, the 20-measure theme and the 32-measure harmonic structure used during the solos. The MJQ plays the theme in a slow, rubato fashion, then takes up a medium swing tempo for the vibes and piano solos. A ritardando at the end of Lewis's solo leads back to the first tempo rubato for the theme's reprise. The structure for the solos consists of 12 measures based partly on the theme, 8 measures on a pedal tone, 4 measures derived from the theme harmonies but in the subdominant key, and 8 measures on a modified boogie-woogie bass figure. Between the solos by Jackson and Lewis is a four-measure interlude, which is the last 8 measures of the theme played quickly. The quartet often plays this piece in concerts, and has recorded it several times since premiering it in 1954. Their finest recording dates from a concert in Gothenburg, Sweden, in April 1960. Everything about this performance is a treasure: the ensemble playing of the somber theme, the flourish on brushes that Kay plays to start Jackson's first solo chorus, the collective build-up of intensity during Jackson's two excellent choruses, the quick reversal of intensity during the transition to Lewis's gentle solo, and the smooth return to the original mood of the theme.

During the 1950s and 1960s the MJQ recorded a series of Lewis's fugues—pieces using extensive imitation of one or more short melodic ideas, or "subjects." This series forms but a small part of the quartet's total œuvre, but is among its most important recordings and is unique in jazz history. *Versailles,* one of the finest in this series, is based on this subject:

The piece alternates between expositions, in which the subject bounces from one melodic instrument to another, and episodes, which are non-imitative sections consisting of 32-measure chordal improvisations and modulating transitions. The outline of this piece, far more than a series of choruses, is as follows:

Exposition 1 (24 measures)—three statements of the subject in the key of C, each answered by a statement in F; after each melodic instrumentalist (vibist, pianist, or bassist) states the subject or its answer, he improvises a countermelody.

Episode 1 (36 measures)—1) Lewis improvises on a 32-measure harmonic structure in *aaba* form (the *a* section uses the chords of *I Got Rhythm*) while Jackson plays the subject as a riff, twice in each *a* section; 2) a 4-measure modulation.

Exposition 2 (16 measures)—two statements of the subject in G, each answered by a statement in C; the counterpoint is similar to that of Exposition 1.

Episode 2 (36 measures)—1) Heath improvises on the 32-measure structure of Episode 1, in the key of G, while Jackson and Lewis accompany (Jackson uses a riff derived from the subject); 2) a modulation.

Exposition 3 (16 measures)—two pairs of subjects and answers in F and B♭.

Episode 3 (36 measures)—1) Jackson improvises on the 32-measure structure of Episode 1, in the key of F; 2) a modulation.

Pedal point (8 measures)—Jackson continues to improvise over a repeated G (the dominant of the main key).

Stretto (5 measures)—three overlapping statements of the subject in C, the tonic key.

Independent polyphony is an essential ingredient in *Versailles,* for Jackson, Lewis, and Heath all play melodic lines constantly throughout the piece. The normal homophonic texture of jazz—improvised melody supported by chordal accompaniment—is all but unknown here. The apparent lack of textural variety could have become monotonous, but in fact the piece has textural variety, thanks primarily to Kay and Heath. In Exposition 1, Kay plays the finger cymbal on beats 2 and 4. Later in the exposition, after Heath has entered, Kay adds a second rhythmic pattern, creating a total of five sound layers. At Episode 1, Kay switches to his ride cymbal and Heath switches into a walking bass line, thereby making clear the change from exposition to episode. At the first modulation Kay returns to the finger cymbal while continuing the ride cymbal, and he marks the beginning of Exposition 2 by changing cymbals and playing a more complex rhythm. Other divisions of the piece are clarified in similar ways.

The piece would mean little if it *sounded* as academic and stuffy as this description makes it appear. But it is actually a wonderful, joyous piece that bubbles and bounces along through 3⅓ minutes. Its complexity is obvious to the ear—these players had a lot to deal with—but, above all, it swings. Further, this performance is a remarkable musical feat. According to Lewis, the Quartet worked only from the subject, written in C and F. He wrote down nothing else of this piece before they recorded—not its structure, keys, chord progressions, or its accompanying melodic lines. *Versailles* may be the most complex and challenging head arrangement in the history of jazz.[11]

Thelonious Monk shared with Lewis an interest in the European tradition,[12] but he had no apparent interest in joining the two traditions. Rather than writing fugues or multisectional compositions as Lewis has done, he wrote mostly 32-measure themes in *aaba* and *aa'* forms and 12-measure blues themes. But for all their formal simplicity, Monk's meticulously crafted pieces typically contain one or more surprises for the unwary player. In the blues *Straight, No Chaser,* the surprise is the evolving nature of the main motive; at first it is a five-note motive starting just before beat 1, then it is a seven-note motive starting just before beat 4, elsewhere it is a four-note motive, and so on. In *Epistrophy,* the surprises are the snaking chromatic harmonies that support multiple repetitions of a quirky five-note motive. In *Evidence* (a *Just You, Just Me* contrafact), it is the unpredictably and widely spaced bits of melody. In *I Mean You,* it is the 31½-measure theme choruses (the solo choruses are 32 measures long). In the ballads *Ask Me Now, Pannonica, Monk's Mood,* and *Ruby, My Dear,* they are the rich and nonconformist harmonic structures. And in the decidedly unorthodox ballad *Crepuscle with Nellie,* the harmony, melody, rhythm, and form are all so unusual that normal improvisation seems entirely inappropriate (indeed, Monk never improvised on this song).

Other than great craftsmanship, there is no single identifying feature common to Monk's pieces. His is a richly varied repertory. He based some pieces almost entirely on one or two short motives—besides *Straight, No Chaser* and *Epistrophy,* there are *Misterioso, Raise Four* (both blues), *Friday the Thirteenth, Criss Cross, I Mean You, Little Rootie Tootie,* and others. Some are less tightly constructed; the classic *'Round Midnight* is a lovely, expansive art song rich in melodic ideas. Some, like *Trinkle Tinkle,* are deliberately awkward and almost impossible to play in unison. But *Blue Monk* is lyrical and simple enough to be sung easily. Some show Monk's impish sense of humor: the impertinent "toot!—toot!—toot!" insistently punctuating *Little Rootie Tootie* (written for Thelonious Jr. when he was a child), the intentional note jumble of *Trinkle Tinkle,* the punning title *Rhythm-a-ning* (his *I Got Rhythm* contrafact). Others, such as the waltz *Ugly Beauty*, are touchingly serious. And some—*Well, You Needn't, Rhythm-a-ning, Bye-Ya*—have a built-in swing that is almost irresistible, especially when his best quartets played them.

Monk had several important saxophonists in his earlier groups, but the one best suited musically and temperamentally to his group was Charlie Rouse (1924–88), who was with Monk from 1959 until 1970. Rouse was not a flashy technician, and he had a unique concept of intonation, but his buoyant tone and fresh melodic ideas are a consistent delight. During those years Monk's bassists were John Ore, then Butch Warren, and finally Larry Gales; his drummers were Frankie Dunlop, then Ben Riley. All made some admirable music with Monk, but the final group—Monk, Rouse, Gales, and Riley—rose to the greatest heights of ensemble playing.

Blue Monk and *Well You Needn't*, from a 1964 jazz-club performance in Los Angeles, are wonderful illustrations of bebop at its most joyous.[13] *Blue Monk* is Monk's simplest, old-time blues melody (even New Orleans street bands play it). The main motive—a four-note chromatic rise in eighth notes—is the melodic springboard for several of Monk's choruses. *Well You Needn't* is a much misplayed theme in *aaba* form (see pp. 116–17). As in *Blue Monk*, the theme is more than a simple frame for solo choruses. It provides the soloists, especially Monk, with material for improvising. There are many motivic and thematic relationships that help unify both performances. Most important, though, these pieces *swing!* No wonder Monk used to get up during Rouse's solo and dance around the stage; it is almost impossible not to move during this music. Several factors contribute to making the rhythm so compelling: Riley's alterations of the swing-eighth ride cymbal rhythm—

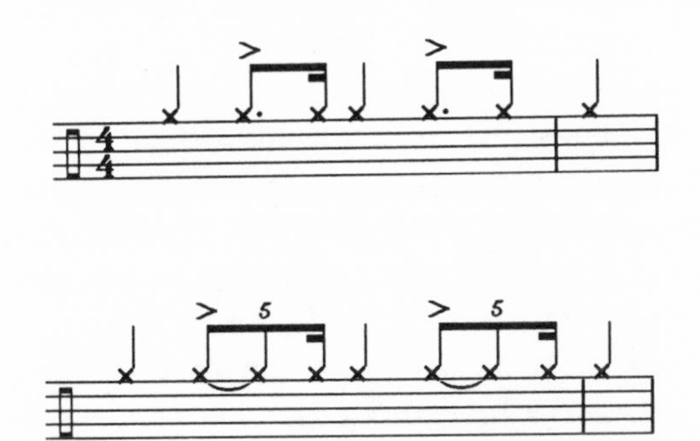

and

—in which the short notes pull even more strongly into beats 1 and 3, the tenor sax/snare-drum dialogue during Rouse's solo, Gales's powerful bass lines, Monk's impish chordal punctuations and melodic ideas, and (most important) all four musicians playing in perfect synchronization.

The classic bebop quintet used by Parker and Gillespie in the mid-1940s—trumpet, saxophone, piano, bass, drums—has been a viable and durable ensemble for bebop. During 1954–56, drummer Max Roach and trumpeter Clifford Brown co-led an excellent quintet with this instrumentation. Brown's front-line partners were all tenor saxophonists: Teddy Edwards briefly, then Harold Land, and finally Sonny Rollins. As with most groups, their repertory included popular standard songs, older jazz standards, and pieces newly composed for the group. Among the latter, Brown made two superlative contributions that quickly became jazz standards: *Daahoud* and *Joy Spring*. But the most stunning group performance of their recorded legacy was *Blues Walk.*[14]

Blues Walk is a simple B♭ blues tune commonly credited to Brown, though it is actually Johnny Richards's *Loose Walk,* first recorded by Sonny Stitt in 1952.[15] Despite the walking suggestion of the title, the tempo is a brisk ♩ = 260. The recording begins with an introduction and two theme choruses. Brown follows with a fine seven-chorus solo, while the rhythm section hums along in perfect, high-spirited synchronization. (The clarity of the recording captures Roach's infectious high-hat backbeats beautifully.) Next comes an eight-chorus solo by Harold Land, who manages to upstage Brown—no easy feat!—with a beautifully executed bebop solo. Pianist Richie Powell, Bud Powell's younger brother, follows with an adequate solo, but coming after the masterful solos by Brown and Land it pales by comparison. We listen patiently, knowing more great moments lay ahead. Brown and Land return with a four-measure unison phrase that sets up Roach for 8 measures. This unison phrase and drum solo return in the next chorus, and then Roach continues an extended solo, developing rhythmic motives in his usual highly musical fashion.

After Roach ends his solo there follows the most extraordinary portion of this excellent performance. For the next five choruses Brown and Land engage in a musical dialogue that has never been surpassed on record so far as I know. The dialogue begins conventionally, with two choruses in which the soloists trade four-measure phrases. These phrases seem unrelated to each other, but there are subtle connections between them. Brown's first phrase unfolds a descent from D to G. Land then picks up the descent by first emphasizing F within a blues-scale context and then making a Parker-like descent of seven notes down to another G. Brown returns, picks up the G an octave higher, then carries it on to F at the start of the next chorus. The baton-passing continues in the next chorus, and on into the third, where they switch the call-and-response pattern to two-measure phrases. Here Land takes Brown's descending phrases and varies them ingeniously. In the fourth chorus the conversation intensifies further, with the players trading one-measure phrases:

Finally, they trade two-*beat* phrases for one chorus:

At the fast tempo of this piece, any further reduction would be independent and simultaneous polyphony; instead, they return to the theme for two triumphant final choruses.

The smoothness with which this musical dialogue proceeds suggests that they preplanned it all, but they did not. In 1983, an earlier version (take 5) of this blues appeared on a compact-disc collection. Here the conversation (which lasts seven choruses instead of five) begins well, but sputters along during the last three choruses. Uncertainty reigns supreme as the players struggle with shorter and shorter phrases, almost all of it different from that used in the final version (take 8).[16]

Shelly Manne, one of bebop's pioneer drummers (see Chapter 2), participated in some of the earliest bebop groups before joining Stan Kenton's big band in 1946. After touring for six years with Kenton he moved to Los Angeles, where he worked with Howard Rumsey's Lighthouse All Stars and then formed a septet of his own. By 1957 he had trimmed the group to a quintet and established a musical partnership with bassist Monty Budwig that would continue intermittently until 1984, when Manne suffered his fatal heart attack. During those years Manne and Budwig became one of bebop's great rhythm teams (see pp. 186, 191).

In September 1959, trumpeter Joe Gordon, tenor saxophonist Richie Kamuca, and pianist Victor Feldman completed Manne's group. They were in excellent form the night that they recorded at San Francisco's Blackhawk jazz club. The highlight of the evening was *Our Delight,* a twelve-minute romp (at ♩ = 265) with Tadd Dameron's *aaba* tune. It features nine choruses by Kamuca, whose style would have blended perfectly into Woody Herman's "Four Brothers" saxophone section of the late 1940s, and shorter solos by Gordon, Feldman, and Budwig. But the real star is Manne, though his only solo is in the bridge of the closing chorus. His energetic yet contained playing supports, complements, and encourages the others brilliantly. His brushwork during Feldman's solo is as ingenious as his stick work earlier. Altogether it is an inspired, high-level performance by this "West Coast" group, consisting of New Yorker Manne, Bostonian Gordon, Philadelphian Kamuca, Londoner Feldman, and Nebraskan Budwig.[17]

Late in 1973 alto saxophonist Phil Woods began using as his rhythm section pianist Mike Melillo, bassist Steve Gilmore, and drummer Bill Goodwin. Originally the core of larger groups, these four men eventually became the Phil Woods Quartet—one of the outstanding ensembles of the 1970s. In 1979 Melillo left and Hal Galper replaced him. In the 1980s the quartet became a quintet, first with trumpeter Tom Harrell, then with trombonist Hal Crook, and most recently with trumpeter Brian Lynch. The empathy and joyous rhythmic verve of these groups led to many recorded treasures; *The Phil Woods Quartet Volume One* (1979), *Live from New York* (1982), *Gratitude* (1986), and *All Bird's*

Children (1990) are four of the best. In the quartet recordings Woods's energetic, dramatic, and creative melodic lines soar effortlessly above a sympathetic, propulsive, and beautifully meshed rhythm section. In the quintet recordings with Harrell there is the added bonus of a remarkable melodic continuity between Woods and Harrell, especially *111-44* from the album *Gratitude,* where the musical conversation recalls the classic *Blues Walk* duet between Clifford Brown and Harold Land. Woods and Hal Crook show a similar rapport in *All Bird's Children.*[18]

Cannonball Adderley's quintets and sextets rank among the top bebop ensembles, largely because of the superb teamwork of the Adderley brothers—alto saxophonist Cannonball and cornetist Nat. The brothers complemented one another perfectly—Cannonball, with the effusive, sweeping melodic flourishes (see pp. 55–59), Nat with the down-home, earthy, hard-swinging melodies (see page 134). Additionally, no front line in the history of jazz played unison lines more perfectly in sync; their instruments seemed to become one (an "alto cornetophone?"). And during 1959–64, the bands included the great rhythm team of bassist Sam Jones and drummer Louis Hayes. The quintet's album done in concert at the Lighthouse in Hermosa Beach, California, captured the group at its best, playing a representative sampling of bebop repertory: a blues (*Big "P"*), a waltz (*Blue Daniel*), a popular standard (*What Is This Thing Called Love?*), and some original compositions by group members (Victor Feldman's *Azule Serape* and Cannonball's folk-tinged "funky" tune, *Sack o' Woe*).[19] Each piece is marvelous.

During most of Art Blakey's career he led various sized groups, calling them all the Jazz Messengers. His taste both as a drummer and as a leader favored aggressive, loud, and hard-driving bebop, and he seemed to impose his taste with every drum and cymbal stroke. He conducted with his playing, molding and shaping the dynamic contours of the themes and of each solo. His sidemen played *his* way or suffered the consequences. Tenor saxophonist Benny Golson said that, when he joined Blakey's group,

> I was playing soft, and mellow, and smooth, and syrupy. By the time I left I was playing another way, because I had to. He would do one of those famous four-bar drum rolls going into the next chorus, and I would completely disappear. He would holler over at me, "Get up out of that hole!" (Lees 1989: Part I, 5)

There are many excellent recorded examples of Blakey's art. *What Know* would be a polished but ordinary sample of bebop without the relentless forward drive of his shuffle beat and crescendos pushing his soloists to give their best. In *Pensativa* his loping, Latin-tinged drum pattern transforms Clare Fischer's beautiful samba into something far different but equally valid. And it is hard to beat the sheer exuberance of the blues *Tell It Like It Is* and the fiery *Free for All.*[20]

In some West African societies the master drummer is more than a good player; operating from a position of high respect, he is a featured player, the shaper of pieces played by the ensemble, and an invaluable repository of musical knowledge that he passes on to others. Art Blakey was such a person in our society. Musicians who have played in his groups often compare the experience with going to school. "Professor" Blakey had many illustrious students in his Jazz Messengers "seminars": trumpeters Clifford Brown, Kenny Dorham, Donald Byrd, Bill Hardman, Lee Morgan, Freddie Hubbard, Chuck Mangione, Woody Shaw, Valery Ponomarev, Wynton Marsalis, Wallace Roney, Terence Blanchard, and Brian Lynch; trombonists Curtis Fuller, Julian Priester, and Steve Turré; saxophonists Lou Donaldson, Hank Mobley, Jackie McLean, Johnny Griffin, Benny Golson, Wayne Shorter, Gary Bartz, David Schnitter, Bobby Watson, Bill Pierce, Branford Marsalis, and Donald Harrison; pianists Horace Silver, Sam Dockery, Bobby Timmons, Walter Davis, Jr., Cedar Walton, John Hicks, Ronnie Matthews, Joanne Brackeen, James Williams, Mulgrew Miller, Benny Green, and Geoff Keezer; and bassists Doug Watkins, Jymie Merritt, Victor Sproles, Reggie Workman, Dennis Irwin, and Charles Fambrough. Few leaders in bebop have had a more imposing list of sidemen.

Horace Silver first arrived in New York City in 1950 as a sideman in the Stan Getz Quartet. For the next several years he was active there, as both a sideman and a leader. In 1953 he began playing with Art Blakey, first in a nonet, then in a quintet that included Clifford Brown, alto saxophonist Lou Donaldson, and bassist Curly Russell. On 21 February 1954, this quintet recorded some crackling performances at Birdland. Three pieces were among Silver's early compositions: *Split Kick* (a *There Will Never Be Another You* contrafact), *Quicksilver* (a *Lover, Come Back to Me* contrafact), and *Mayreh* (an *All God's Children Got Rhythm* contrafact).[21] On these and other bebop standards the group's enthusiasm is obvious; buoyed up by Blakey's compelling drumming, Silver, Brown, and Donaldson play some excellent solos. The group did not remain intact long, for Brown soon left to form his historic partnership with Max Roach. But Silver and Blakey remained associates until 1956 in various Jazz Messengers quintets.

In 1957, Silver ceased all sideman activities—he had recorded prolifically for various leaders since moving to New York—and has performed and recorded exclusively as a leader ever since, mostly for Blue Note (1952–79), then for his own recording companies Silveto and Emerald (the 1980s), and most recently for Columbia. During this long period Silver's basic musical esthetics have not changed significantly. His usual ensemble is the quintet, sometimes augmented by one or more singers. With few exceptions his groups only play and record his music. Unlike Monk, who returned again and again to several of his favorite compositions, Silver continues to record new pieces, and rarely re-records anything.

Some of Silver's early pieces fit the typical bebop mold of the late 1940s and early 1950s; they are blues melodies (*Doodlin'* and *Opus de Funk*) or melodically active contrafacts of older popular songs (*Stop Time,* based on *Lazy River,* plus those listed earlier). But soon he abandoned this approach and explored other compositional avenues.[22] Almost from the beginning there have been some songs with a gospel or folk blues flavor. In the gospel style are *The Preacher* of 1955 (whose melody and harmony derive from *Show Me the Way to Go Home*), and *Don't Dwell on Your Problems, Let the Music Heal Ya,* and *Hangin' Loose* thirty or more years later. In the folk blues style are *Doodlin'* and *Señor Blues* in the 1950s and *Everything's Gonna Be Alright* in the 1980s. In many pieces his explorations of the minor-pentatonic and blues scales almost ensure the folk connection.

But for every down-home, "funky" tune there is one or more that in no way fits the "funky" stereotype. There are ballads, most notably *Lonely Woman* (recorded in a piano trio setting), and the touchingly beautiful *Peace* (especially the version with Andy Bey singing Silver's sensitive lyrics). There is a great array of Latin-tinged pieces, including the bolero *Moon Rays,* the bossa nova *Gregory Is Here,* and the samba *Time and Effort.* His harmonic vocabulary in *Melancholy Mood, Barbara,* and others is far removed from the folk traditions, as is the $\frac{7}{4}$ meter of *Perseverance and Endurance.* There are pieces with phrase structures unlike those of the 12- and 24-measure blues and the 16- and 32-measure song forms of gospel music. The *Outlaw* has phrases of 7, 6, 7, 6, 10, and 18 measures (*ababcd* form); *Nineteen Bars* has phrases of 7, 8, and 4 measures; *New York Lament* is in the standard *aaba* form, but the *a* sections are 22 measures each; *Activation* is also in *aaba* form, with phrases of 9, 9, 14, and 9 measures.

The most obvious change taking place in Silver's artistic evolution appears not in music but in the words. In 1968, lyrics written by Silver appeared on the liner notes for *Serenade to a Soul Sister,* and though no one sang those lyrics on the album, their appearance signaled a new-found interest. Since 1970, he has written lyrics and added singers to his ensembles for most of his recorded compositions. These pieces focus on extramusical topics, especially on the improvement or healing of the mind, body, and soul: *Wipe Away the Evil, Won't You Open Up Your Senses, Soul Searchin', How Much Does Matter Really Matter, Who Has the Answer?, Freeing My Mind, Inner Feelings, Optimism, Expansion, Accepting Responsibility, Don't Dwell on Your Problems, There's No Need to Struggle, Music to Ease Your Disease,* and others.

Few jazz musicians prepare more thoroughly for recording than Silver. Typically he works on these pieces for months or more before bringing them to his group; the group then plays them in concerts repeatedly before recording them. Thus, many performance details are polished before recording begins, and the ensemble playing is often of a high order.

Some of Silver's best compositions are those in which unifying accompanimental textures hold the entire performance together. *Nica's Dream* is a case in point. The trumpet melody, one of the great themes in jazz literature, is a 64-measure song in *aaba* form. The accompaniment for the *a* sections is in a Latin style based on the rhythm | ♩. ♪‿♩ ♩ |—one of Silver's favorite patterns. In the bridge the accompaniment alternates between backbeat chordal punctuations and four-beat swing. During the solos the rhythm section maintains the same accompanimental textures, which both clarify the form and maintain the theme's original moods and textures. A recurring eight-measure interlude is a further unifying device. With such a structured arrangement coherence is almost assured. The soloists in the 1960 recording, buoyed by an expert rhythm section and guided by the structural signposts Silver has provided, play fine solos. The result is a memorable performance.

Silver's most famous composition is *Song for My Father,* and his 1964 recording of it is his most famous recording. The moderate-tempo theme is 24 measures long, in *aab* form. It contains only four chords: $F\text{MI}^{9}$-$E^{\flat 9}$-$D^{\flat 9}$-C^{9}. The piece uses even eighth notes throughout, not swing eighths. Two aspects of the accompaniment are just as important as the melody itself. First, the bassist plays only roots and fifths of the chords, in the rhythm | ♩. ♪ ♩. ♪ |. Second, the rhythm section breaks at measure 6 in every section. Both the bass pattern and the one-measure break occur consistently during the solos as well as the theme choruses, and provide an extra degree of continuity that helps tie the whole performance together. (Silver is fond of such accompanimental procedures; his bass players seldom simply walk, but usually play set patterns throughout each piece.) A chordal punctuation based on this rhythm——that Silver plays in the seventh and eighth measures of several sections provides an additional thread of continuity.

After the front line plays the theme twice, Silver plays a three-chorus solo that is a model of tight construction. He begins with a left-hand comping pattern that largely duplicates the bassist's pattern: . This pattern dominates his first eight measures, and returns intermittently throughout the solo. Additionally, he uses some recurring melodic patterns, the most obvious being this one:

This figure recurs three times, always in measure 6 of a section. Variations, based on the E♭-D-C descent, occur also; here are three, the last using the inversion, C-D-E♭:

The first and third of these variants also contain lower-neighbor embellishments of scale degree 5 (C in these examples). These embellishments, which stem from early jazz pianists' imitations of blue notes, form one more recurring thread of continuity in this solo. There is an overall crescendo of intensity, beginning with the sparse melody of the first eight measures, then the more active second and third eight-measure phrases, the syncopated sixteenths of the second and third choruses, and finally the louder block chords of the final eight measures. It is a classic example of how Silver builds effective larger structures with simple musical means.

After Silver finishes, tenor saxophonist Joe Henderson takes over, to play an extremely dramatic solo. He begins almost as simply as Silver did, even using a figure derived from one that Silver had played earlier. He maintains a low level of activity and range until his surging pentatonic run in the second eight measures. From then on the drama increases, with more short, syncopated notes, higher notes, and especially notes played with an intense tone quality. All the while Silver's block-chord comping offers him solid support. When Henderson ends his second chorus his intensity level is too high to make an immediate return of the theme acceptable. So he adds ten extra measures on the tonic chord, and gradually reduces the excitement level. Then he and trumpeter Carmell Jones reprise the theme. A short coda closes this famous performance.[23]

Horace Silver's quintets, the Modern Jazz Quartet, and other great ensembles usually play well-rehearsed pieces and refine their ensemble playing over months or years of performances. But Miles Davis preferred to put together performances and recordings in the most casual way possible, letting things fall into place spontaneously. Some of his classic recordings stem from his group's first moments with the pieces, moments when written music was notably absent or incomplete. Bill Evans, in his liner notes for the album *Kind of Blue* (see below), underscored this point:

> Miles conceived these settings only hours before the recording dates and arrived with sketches which indicated to the group what was to be played. Therefore, you will hear something close to pure spontaneity in these performances. The group had never played these pieces prior to the recordings and I think without exception the first complete performance of each was a "take."

Davis led at least three classic ensembles: the quintet of 1955–56, with John Coltrane, Red Garland, Paul Chambers, and Philly Joe Jones; the sextet of 1958–59, with Cannonball Adderley, Coltrane, Bill Evans or Wynton Kelly, Chambers, and Jones or Jimmy Cobb; and the quintet of 1963–68, with George Coleman and later Wayne Shorter, Herbie Hancock, Ron Carter, and Tony Williams. Each group reached the highest levels of ensemble skill not through preplanning, but through great sensitivity to one another's ideas and impulses. They usually avoided unison theme statements, transitions, and endings, unless they were easy to learn and remember (such as the two-chord riff of *So What,* the background pattern for *All Blues,* or the generally known transition in *'Round Midnight*), so extensive writing and rehearsing was unnecessary.

A typical case is his 1956 recording of the old popular song *Bye, Bye, Blackbird.* It begins with Red Garland alone, playing an introduction. Chambers and Jones enter to start the theme chorus while Davis skates casually but expressively over their two-beat accompaniment, and Garland reacts sympathetically to Davis's phrases. The theme ends with the ubiquitous two-measure break, and Davis continues on with his solo, now more aggressively accompanied in four by his perfectly synchronized rhythm section. When he finishes, Coltrane enters for the first time; his solo lasts for two choruses, and then Garland solos for one and a half choruses. Davis returns to remind us of the theme, and then leads the rhythm section into a universally understood coda. The preparation for this recording could have been no more than Davis announcing the title and key, and counting off the tempo. Yet because each player knew the piece, knew from previous experience what kinds of rhythms and textures to play, and knew each other's musical habits, the performance came together brilliantly.[24]

The great sextet of 1959 reached an historical pinnacle of ensemble jazz in the album *Kind of Blue* (see pp. 57–59, 96–97, 120–23, 160) by using the same casual approach to performance organization. The blues pieces *Freddie Freeloader* and *All Blues,* the harmonically expressive *Blue in Green,* the scale-oriented *So What,* and the quietly magnificent *Flamenco Sketches* are as central to the jazz tradition as Beethoven's string quartets are to the European chamber music tradition. Every student of jazz should own this recording.

Flamenco Sketches is one of several in the Davis Sextet repertory that seems to provide a scalar rather than chordal basis for improvisation. In this piece, which has no theme, there are five segments in each chorus: one each in C major, A♭ Mixolydian, B♭ major, D "Spanish Phrygian," and G Dorian.[25] Each solo is a different length because the players decided spontaneously how long to stay in each tonality. The outline of the piece is as follows:

Evans introduction:	4 measures
Davis solo:	4 + 4 + 4 + 8 + 4 measures
Coltrane solo:	4 + 4 + 4 + 8 + 5 measures
Adderley solo:	8 + 4 + 8 + 8 + 4 measures
Evans solo:	8 + 4 + 8 + 4 + 4 measures
Davis solo:	4 + 4 + 4 + 8 + 2 measures

The piece has a quiet, impressionistic character throughout, due in large measure to the bittersweet harmonies and gentle touch used by Evans, both in his comping for others and in his Ravel-like solo. Cobb stays on brushes and Chambers plays a quiet, supportive part, mostly on beats 1 and 4. Davis plays two expressive solos, using his Harmon mute, and the saxophonists maintain the mood beautifully in their solos.

Adderley's solo in *Flamenco Sketches,* discussed in Chapter 4, is filled with stunningly beautiful improvised melodies and exhibits a sensitive, almost telepathic teamwork among Adderley and his accompanists. The transition to each new key (or chord) occurs flawlessly, whether after four or eight measures in the previous key. Admittedly, eye contact or other visual cues might have helped them make these transitions. But other elements could not have been cued visually. The rhythmic give and take, at first between Adderley and Evans (measures 6 and 7) and then among all four players (measures 8–12 and 18–31), is subtle but unmistakable. The gentle cat-and-mouse game they play with the meter, shifting between $\frac{4}{4}$ and $\frac{8}{8}$, is group improvisation at its subtle best; it makes Adderley's solo the rhythmic high point of the performance.[26]

A year after recording *Kind of Blue,* John Coltrane left the Davis Sextet and formed his own quartet, whose personnel eventually became pianist McCoy Tyner, bassist Jimmy Garrison, and drummer Elvin Jones. This group played an important role in the movement toward an expanded bebop vocabulary. Eventually the group ventured beyond the bebop idiom and made important contributions to the action jazz ("free" jazz) idiom as well.

Coltrane recorded a varied repertory in the recording studios, but settled on a much more limited repertory for concert performances. The repertory changed somewhat from concert tour to concert tour, but typically included most or all of these pieces: *Impressions,* a fast *So What* contrafact; *Mr. P. C.,* Coltrane's blues tribute to Paul Chambers; *Naima,* Coltrane's gentle tribute to this first wife; *Afro-Blue,* a triple-meter Latin jazz standard by Mongo Santamaria; and the popular song standards *Ev'ry Time We Say Goodbye, I Want to Talk about You, Inchworm,* and *My Favorite Things.* His favorite pieces on this short list appear to have been *Impressions* and *My Favorite Things;* David Wild's Coltrane discography lists over two dozen concert and television performances of each. Both pieces are direct outgrowths of his years with Miles Davis, when the players were moving away from chord-based improvising. The melody of *Impressions* is in the Dorian scale;[27] the other piece requires a little longer explanation.

My Favorite Things was originally a bright, bouncy waltz in *aaab* form in the 1959 Rodgers and Hammerstein musical *The Sound of Music.* Coltrane kept it in triple meter, but used its 16-measure sections to frame scale-based improvisations of indefinite lengths. A performance consisted of three solos—one by Coltrane, one by Tyner, and another by Coltrane—bounded by an introduction and coda of varying lengths. Because of the open-ended interpolations, different recorded performances vary in length from ten to fifty-six minutes.

The performance of 26 October 1963 illustrates well this distinctive structure:[28]

Introduction—44 measures.
Part 1 (Coltrane)—280 measures
- *a* section of theme—16 measures
- improvisation on E Dorian—64
- *a* section of theme—16
- improvisation on E major—136
- *aa* sections of theme—32
- improvisation on E Dorian—16

Part 2 (Tyner)—480 measures
- *a* section of theme—16
- improvisation on E Dorian—184
- *a* section of theme—16
- improvisation on E major—128
- *a* section of theme—16
- improvisation on E Dorian—104
- *a* section of theme—16

Part 3 (Coltrane)—512 measures
- improvisation on E Dorian—16
- *a* section of theme—16
- improvisation on E Dorian—192
- *a* section of theme—16
- improvisation on E major—240
- *ab* sections of theme—32

Coda—33 measures

This 21½-minute performance is held together partly by the ritornello-like theme segments, but also by motivic developments in the improvised sections. Coltrane weaves a fascinating tapestry of melodic threads, unfettered by any predetermined structural boundaries. In these performances the group clearly is redefining the traditional roles and relationships of jazz-group members. Garrison avoids playing walking bass lines, and instead concentrates on rhythmically more varied melodic lines. Tyner supplies an almost constant chordal barrage during Coltrane's lengthy solos. And Jones aggressively interacts with the soloists, creating energetic and dense textures with them. He seems to play drum solos during the sax and piano solos, but he is actually entering into a lively dialogue with them. The result is the musical equivalent of animated and excitable old friends conversing, interrupting one another, yet interacting and exchanging thoughts. It is a fascinating and brilliantly executed musical experience, one far removed from earlier bebop.

Bassist Charles Mingus led varied groups during his career—trios, quartets, quintets, and sextets mostly, but occasionally larger groups as well. Even though some groups remained intact for one or more years, there was often a deliberate disorder to many of their performances. Mingus loved the jumbled sound of collective improvisation, both in the New Orleans jazz spirit (*Folk forms No. 1*) and in the more radical style bordering on action jazz that emerged briefly in many pieces. He also preferred a not-quite-unison playing of melodic lines, and, perhaps to ensure that effect, sometimes taught his group his new pieces by ear rather than through notation. Sometimes the jumbled sound had a programmatic significance (*Haitian Fight Song, Wednesday Night Prayer Meeting, Pithycanthropus Erectus*); other times it probably was there just because he liked it. Not everything was jumbled, however. The band played his *Fables of Faubus* (inspired by the Governor of Arkansas who defied the Supreme Court decision concerning racial integration of public schools) cleanly and precisely. Despite a title that has lost much emotional impact over the years, the piece is one of his most effective musically.

Mingus was a powerful man with a violent temper, who played his instrument with great power and energy and wrote correspondingly energetic, often dissonant pieces. Yet he had a tender side. His mournful elegy for Lester Young, *Goodbye Pork Pie Hat,* is a touching and heartfelt blues. Another elegy, *Duke Ellington's Sound of Love,* is a gorgeous ballad that incorporates some of Ellington's and Strayhorn's luscious harmonic touches. He had many musical tastes, including a genuine love of African-American folk music, the essence of which he captured in some of his bass solos and themes. The black gospel tradition sings out in *Better Git It in Your Soul.* The New Orleans tradition sings out with equal clarity in *Pussy Cat Dues.* His deep admiration for Ellington's music is unmistakable in many of his pieces in addition to *Duke Ellington's Sound of Love;* for example, it governs much of the texture, instrumentation, melody, and performance style in his extended composition for eleven instruments, *The Black Saint and Sinner Lady.* He also loved straightforward bebop, as in *No Private Income Blues.*

Mingus's rhythm partner during his most creative years was drummer Dannie Richmond. The musical communication between the two men was so remarkable that at times it even confounded the other sidemen. That communication was a key to the importance of Mingus's groups, for underpinning the sloppy unisons and rough-edged collective improvisation was a compelling swing generated with total assurance by Mingus and Richmond. That uncanny rhythm team also controlled the sudden shifts of textures and dynamics that characterize *Folk Forms No. 1* and other pieces.[29]

Big bands, which dominated the swing era, are much rarer in bebop; but in the 1940s there was a handful—led by Dizzy Gillespie, Woody Herman, Stan Kenton, and a few others—that adopted the complexities of the new jazz language (see pp. 20–22). Herman and Kenton continued their careers as full-time road-band leaders until the ends of their lives, and made many fine contributions to the recorded literature of bebop. The Ellington and Basie bands (especially the latter) gradually evolved into "swibop" bands, finding magnificent blends of elements from the two styles. In the 1960s drum virtuoso Buddy Rich launched a successful road band that worked until 1974, and then he started another band in the 1980s. In Europe, American drummer Kenny Clarke and Belgian composer/pianist Francy Boland co-led a fine band of American and European musicians during the 1960s and early 1970s. In Canada, valve trombonist Rob McConnell formed his reedless Boss Brass big band in 1968; by 1971 he had added reeds but retained the original name for his excellent and unusually large band. These and other similar groups proved repeatedly that bebop is not necessarily small-group jazz.

The typical bebop big band differs little in instrumentation from the typical swing band. For both styles, three to five trumpets, three or four trombones, five woodwind players, and three or four rhythm players are the norm. Also, much writing in both styles uses similar orchestration procedures. Arrangers in both idioms typically used 1) trumpet-trombone octave doubling of melody against saxophone chords, 2) saxophone octave doubling of melody against brass choral punctuations, 3) trumpet unison melody against trombone chordal punctuations, 4) trombone unison melody against trumpet-saxophone chords, 5) all winds in block chords. The riff, that much-used swing-era melodic device, remains valid in bebop. And blues, plus *I Got Rhythm* and other 32-measure songs, make up much of both repertories.

The differences between the two big-band styles include the increased level of dissonance in bebop arrangements, the new melodic formulas and articulation habits of bebop, and the striking rhythmic changes that took place in bebop rhythm sections. Woodwind sections are different, too. Bebop saxophone players seldom double on clarinet (popular in swing bands), but the lead alto player often doubles on soprano saxophone (all but unknown in swing bands), and one or more saxophonists may double on flute (also rare in swing). Finally, the range of bebop trumpet parts has expanded upward, not just for the lead players, but for the entire section. This increased range, which seemed so challenging in the 1940s, is now the norm, and players have adjusted to it.

But while the music played by bebop big bands represents a continuation of a tradition established in the 1920s through early 1940s, the working conditions have been radically different. The prime employment conditions for big bands ended in the early 1940s, and have never returned. Most big-band performances today are concerts, not dances, and the concert opportunities are few compared with the dance-job opportunities of the 1930s. Although Herman, Kenton, and Rich were able to perform year-round and provide a living for their sidemen, most bebop big bands have been part-time ensembles. Some—such as Quincy Jones's various bands—consist of studio men hired for an occasional recording session. More often, however, they are groups that perform publicly a few times each year and seldom if ever leave their home region. In spite of their part-time status, however, Bill Potts's Orchestra, the Blue Wisp Big Band, the Capp-Pierce Juggernaut, and the bands led by Bill Holman, Bob Florence, Louie Bellson, Gerald Wilson, and others are first-rate and distinctive.

Woody Herman (1913–87), though a swing-style clarinetist and saxophonist, led bebop bands almost continuously from the mid-1940s until the year he died. And though he kept some popular arrangements from the early years in his bands' repertoires, most notably the classic *Four Brothers* (see pp. 21–22), he continually added new arrangements, many contributed by his sidemen. In the early 1960s his band book included pieces by Horace Silver, Herbie Hancock, and others. His bands played with great spirit and precision, especially on the concert albums recorded in Los Angeles.[30] In the late 1960s Herman found some interesting arrangements of popular rock tunes. In the early 1970s, Alan Broadbent, Bill Stapleton, and others contributed some excellent arrangements of newer jazz standards and some ambitious original compositions. By this time the rhythm section included electric keyboard and electric bass, and the repertoire included some fusion-influenced pieces. In the 1980s Herman, by then old enough to be his sidemen's grandfather, continued encouraging them to arrange and compose new material. In those last years, as illness, financial difficulties, and his wife's death threatened to overpower him, he seemed to draw strength from the energy, enthusiasm, and musical discipline his sidemen displayed.

During Herman's final illness the band continued touring, led by saxophonist Frank Tiberi (born in 1928), who joined the band in 1969. Tiberi continues to lead the band, one of the world's few remaining road bands. Their repertoire includes the swing-style *Woodchopper's Ball* (Herman's 1939 hit), many bebop standards, some fusion pieces, some adaptations of European style concert pieces, and Stravinsky's *Ebony Concerto,* a non-jazz piece written for the 1946 Herman band.

The swing band led by William "Count" Basie (1904–84) also showed traces of the bebop world in the 1940s, primarily during 1945–46, when the trombone section included J. J. Johnson. But the band he formed in the early 1950s, after a short period devoted to leading smaller groups, started a dramatic new phase in Basie's life. Before the decade was over the band book included many new arrangements by Ernie Wilkins, Frank Foster, Quincy Jones, and Neal Hefti, and the band featured bebop soloists Thad Jones, Joe Newman, Benny Powell, Frank Foster, and Frank Wess.

The Basie band never became a full-fledged bebop band. For one thing, Basie maintained the irresistible, straightforward swing groove he established in the late 1930s with his famous rhythm partners Freddie Green, Walter Page, and Jo Jones. Later bassists amplified their instruments and later drummers added bebop licks to their readings of the arrangements, but Basie's spare piano fills and Green's quiet four-beat strumming kept the rhythm section anchored to its swing-era beginnings. In addition, most of the arrangements contained elements from the older style—riffs, uncomplicated melodies, and clear antiphonal effects between the brasses and reeds. The writing was in an updated swing idiom, which served as a backdrop equally well for Marshall Royal's swing-style solos and Frank Foster's bebop-style solos. The band in the late 1950s and early 1960s had a unique sound, one of the classic musical styles of the twentieth century. Basie maintained the integrity of this hybrid style until just weeks before his death. Since then Thad Jones, followed by Frank Foster, alumni of the 1950s band, have maintained the Basie institution with dedication.[31]

In December 1965, ex-Basie-band trumpeter and arranger Thad Jones (1923–86) and famed big-band drummer Mel Lewis formed a rehearsal band of studio musicians to play Monday Nights at the Village Vanguard. During the following years a series of impressive recordings and occasional national and international tours established the band's importance. Jones's skillful writing attracted fine players, who in turn developed a cohesive sound rivaling that of the Basie band. During those fertile years Jones's *Kids Are Pretty People, The Groove Merchant, A Child Is Born,* and other pieces written for and recorded by the band became jazz standards.[32] Jones left the band in 1979 and moved to Europe. (He returned to the U.S. in 1984 to lead the Basie band.) Mel Lewis then continued leading the band, eventually bringing in Bob Brookmeyer as chief arranger, until shortly before his death.

The most venerated big-band leader / composer / arranger in bebop is Gil Evans (1912–88). Writers and players often compare his best work to that of Duke Ellington. Ellington clearly inspired some aspects of Evans's strongly personal style, and his sidemen treasured their opportunities to play his music much as Ellington's sidemen treasured their opportunities. Yet, unlike Ellington, he could not offer his sidemen a living wage, only an occasional recording session and a few scattered night-club engagements and tours.

Evans's arranging career began in the 1930s. By at least the middle 1940s he had developed a distinctive harmonic and orchestral style, as evidenced by his arrangements for Claude Thornhill's big band. Part of that orchestral style stemmed from Thornhill's unusual brass section: three trumpets, two trombones, two French horns, and tuba. This instrumentation, in turn, influenced the Miles Davis nonet of 1948–50, for which Evans arranged *Boplicity, Moon Dreams,* and *Darn That Dream* (see pp. 23–24).

Those early arrangements, an album for singer Helen Merrill in 1956,[33] and a limited array of other arrangements (see Fox 1959: 100ff.) were preambles to three great album collaborations with Miles Davis during 1957–60: *Miles Ahead, Porgy and Bess,* and *Sketches of Spain.* The first of these is the most significant, for in it Evans established the basic instrumentation and concerto-like relationship between soloist and band that he would use in the later albums. The instrumentation is a variant of the resources he used in the Thornhill and Davis nonet recordings, but with more brasses and a more varied woodwind palette. The album's subtitle, *Miles + 19,* indicates the size of the ensemble: five trumpets, three trombones, bass trombone, three French horns, tuba, four woodwinds, bass, drums. Significantly, there is no pianist here, or on the subsequent recordings. Evans, a pianist himself, knew well that big-band pianists spend much of their time avoiding clashes with the winds, and comparatively little time playing. Further, the band needed him more as a conductor than as a pianist.

Miles Ahead contains ten pieces by nine different jazz and popular song writers (including Evans and Davis) and one by European composer Léo Delibes. Evans linked most of them by simple means, such as a single sustained note played pianissimo. This linking, plus Evans's consistently colorful wind textures and Davis's poignant flügelhorn playing, creates the effect that these disparate pieces actually form a single large work by one composer. And in a sense they are, for just as composers in the European concert-music tradition have taken folk songs and shaped them into symphonies, ballets, and song cycles, so Evans has created a major work—a beautiful through-composed concerto for Davis and the wind ensemble—out of some short songs. The beauties of the original songs are meaningful, especially Dave Brubeck's *The Duke,* J. J. Johnson's *Lament,* and Kurt Weill's *My Ship.* But to savor the delicious blend of French horns, flutes, bass clarinet, muted trumpets, muted trombones, and Davis's matchless flügelhorn tone in Evans's reharmonization of *My Ship* is to hear music of a higher order than Weill's lovely, simple song.

There are flawed moments in *Miles Ahead,* and in *Porgy and Bess* as well. Evans did not write the same kinds of phrases nor use the same voice leading that most writers used. His music was difficult, even for the top studio players hired for the sessions. And his varied instrumental blends called for unusual efforts by the players to achieve proper dynamic balances and unity of phrasing. The recording schedule provided insufficient rehearsal time for polishing the most difficult passages. But in the end it does not matter, for the music is wonderful. The music is excellent in *Sketches of Spain,* too, and the performances are more polished, for, according to Carr (1982: 114), the musicians had fifteen three-hour sessions to record the Spanish album. The previous albums were completed in only four sessions each.

While collaborating with Davis, Evans wrote other classic works for big band. In 1958–59 he recorded two albums of jazz standards drawn from the New Orleans, swing, and early bebop traditions, with Cannonball Adderley as featured soloist in the first. In the 1960s he wrote some gems equal to anything he did with Davis. His album *Out of the Cool* contains the exceedingly simple but effective *La Nevada* and the somber *Where Flamingos Fly.* The latter, a vehicle for trombonist Jimmy Knepper, brings to mind the most expressive *Sketches of Spain* pieces. And *The Individualism of Gil Evans* contains the exquisite and haunting *The Barbara Song* and the bitingly dissonant and dramatic *El Toreador.* In these albums Evans plays piano with great expressivity and a composer's sense of the right texture and phrase for the moment.[34]

Toshiko Akiyoshi (born in 1929) is a big-band composer with a musical voice rivaling Evans's. She began writing for big bands in the 1960s, but was known primarily as a small-group pianist until she and her husband, flutist-tenor saxophonist Lew Tabackin, formed a rehearsal band of Los Angeles studio musicians in 1973. Like the Thad Jones/Mel Lewis band a continent away, this band soon had a series of important recordings to its credit and was performing several concerts per month. In the early 1980s Akiyoshi and Tabackin moved to New York City, where they formed another big band.

Kogun, one of Akiyoshi's most stunning works, is one of several pieces in which she enriches the vocabulary of big-band bebop with elements of Japanese traditional music.[35] There are two recordings, one from 1974 and a better one from 1976. For the second recording (done in Osaka), the band is joined by two Noh drummers, who also intone the traditional vocables, or *kakegoe,* that are an essential part of the Noh drama. Their drum strokes and guttural vocables punctuate nonmetrically the opening and closing portions of the piece. The opening and closing pentatonic melody sounds much like a Japanese folk song. Akiyoshi has the woodwinds play it in a near-unison that evokes the heterophony of the ancient Imperial court music, *gagaku.* It is an unforgettable effect.

The central portion of the piece begins with a solo by Akiyoshi in E Dorian; it is mostly pentatonic (E-G-A-B-D), and is unified by a repeated-note motive. The band then enters with bits of the opening woodwind theme and some repeated notes of its own. Then Tabackin, one of the jazz world's greatest flutists, plays a beautiful and dramatic cadenza, in which he maintains the pentatonicism of the piece. The band and the Noh musicians re-enter to complete the large ABA structure. In the end, Tabackin and one of the Noh singers meet on an E, which they sustain until the final punctuating stroke of the Noh drum.

In the 1950s a few jazz and non-jazz composers wrote some works that attempted to blend jazz and European concert music and create a new music. Gunther Schuller, a major composer and central figure in the idiom, labeled it third-stream music. Most of it joined a jazz group with a chamber ensemble or orchestra. Often the jazz musicians had to deal with structures, harmonies, and melodies that were foreign to their idiom, while the European-oriented musicians had to cope with equally unfamiliar rhythms, phrasings, and articulations. The results sometimes made the listener wish the jazz group would simply play a jazz tune alone, and then let the chamber group or orchestra play some Bartók. But occasionally efforts to fuse the two traditions have worked. John Lewis's reworking of *God Rest You Merry Gentlemen,* which he titled *England's Carol,* some of his fugues (discussed above), and Jimmy Giuffre's *Tangents in Jazz* and *Piece for Clarinet and String Orchestra* are among the early successes in the genre.[36]

Because the performance practices of jazz and of European concert music are substantially different, the composer must write with extraordinary care unless some or all of the players are equally skilled in both traditions. In the 1950s perhaps the only bebop musician who was truly bimusical was pianist André Previn, and he did not participate in third-stream music. A few more are now in that category, including pianists Chick Corea and Keith Jarrett, flutist Hubert Laws, saxophonist Branford Marsalis and his brother, trumpeter Wynton Marsalis, bassists Eddie Gomez, George Mraz, and John Clayton, and clarinetist Eddie Daniels.

Eddie Daniels's skill as a clarinetist is unsurpassed in jazz, and because of his command of both jazz and European concert music, he is as comfortable playing the Mozart Clarinet Quintet or the Poulenc Clarinet Sonata as he is playing *Stella by Starlight* or *Cherokee.* With these skills he is well equipped with deal with music that combines the two traditions, as he shows spectacularly in his recording of *Solfeggietto / Metamorphosis.*[37]

Solfeggietto, Carl Philipp Emanuel Bach's most famous composition, was originally a short keyboard toccata. Arranger Jorge Calendrelli transposed it from C minor to D minor, and reworked it into a demanding bimusical showpiece for Daniels, orchestra, and jazz rhythm section. The beginning section, except for some fleshing out of Bach's implied harmonies, and two two-measure cuts, presents a complete statement of Bach's composition, played faster and more elegantly than Bach could have imagined in the eighteenth century. Daniels gives no hint of his jazz abilities here; his tone, phrasing, attacks, and ultralight vibrato are those of the finest orchestral player. Then an orchestral transition leads to a more relaxed middle section. Here the jazz starts; it is an improvisation on Bach's harmonic structure, embellished and extended, and is as dazzling in its way as was the opening theme statement. Daniels exerts total control over the instrument and over the crescendo and decrescendo of the musical drama. He swoops effortlessly over the clarinet's range, imparting a rich and full tone to every note—including the highest B♭, an extremely difficult note to play. Above all, the music swings lightly but convincingly. Next follows a second transition and a return to the original piece, this time with the cuts restored. Daniels reminds us of his jazz vocabulary with a set of superb out-of-tempo cadenzas, and then the piece concludes.

To close this chapter I want to discuss two unorthodox bebop ensembles, both involving pianist Chick Corea.

Corea sometimes teams with vibraphonist Gary Burton to perform duo recitals. Beside having great technical skills and the ability to play great ensemble jazz, both men are excellent solo players. So when they play duets they can create uncommonly dense polyphonic textures. *La Fiesta,* Corea's famous Latin-tinged composition, is one of their finest duets. Their ten-minute performance, recorded in 1978, begins with an extended free-meter introduction in which the players take turns improvising. The main body of the piece contains two themes in $\frac{6}{4}$, one based on an eight-measure ostinato in E "Spanish Phrygian," the other in A major. After they play the themes they return to the ostinato for an extended solo by Corea, with Burton accompanying. A return to the themes serves as a ritornello. Then Burton solos while Corea comps energetically. A return to the themes leads to a collectively improvised coda. It is a joyous performance, thoroughly in the spirit of the song title.

Corea's versatility has enabled him to perform in several different jazz contexts during his career: Latin jazz, bebop, action jazz, and fusion. He is most famous for his fusion jazz, principally the various Return to Forever groups of the 1970s and his Elektric Band of the 1980s and 1990s. But often he switches from bebop to fusion and back again. In fact, the core of his original Elektric Band—himself on keyboards plus electric bassist John Patitucci and drummer Dave Weckl—also played as a piano trio under the name of the Akoustic Band. And on their debut album the Elektric Band included a wonderful bebop trio piece, *Got a Match?*

The instrumentation hardly suggests bebop. Corea plays two stacks of Yamaha synthesizers linked together via MIDI (Musical Instrument Digital Interface);[38] Patitucci plays a six-string electric bass; and Weckl plays a set of drums tuned far differently from the way bebop drummers usually tune their sets. But the piece is a 16-measure bebop-style theme built on typical bebop chordal relationships, and during much of Corea's solo Patitucci plays walking bass lines and Weckl plays variations on the standard ride-cymbal pattern. The performance is a glorious example of ensemble playing. Corea and Patitucci, playing an octave apart, perform almost flawlessly the lively, complex theme and two even more complex secondary themes. Corea and Weckl play carefully worked out rhythm punctuations during Patittucci's solo. And throughout, the rhythmic drive is infectious.[39]

The preceding discussions barely scratch the suraface of the important bebop ensembles and their repertories. Each ensemble deserves a separate, large chapter, or a complete book. I hope these few representative ensembles and pieces are sufficient to show that from the perspective of the early 1990s, the richness of this great musical tradition seems boundless.

CHAPTER ELEVEN

Younger Masters

You don't see no Charlie Parkers coming along. (Dizzy Gillespie[1])

With young people like these to sustain [jazz], you don't have to worry about the future. (Frank Wess[2])

Since the advent of bebop, two other major jazz styles have emerged. The first, variously called "the new thing," "avant-garde jazz," "free jazz," "action jazz," and other labels, began in the late 1950s. Its early practitioners, including groups led by pianist Cecil Taylor and saxophonist Ornette Coleman, challenged many of the melodic, harmonic, rhythmic, textural, and formal concepts of bebop. Action jazz represented a bigger stylistic change than bebop had represented twenty years earlier; not surprisingly it was greeted with widespread confusion and alienation. There continues to be a worldwide network of skilled practitioners and supporters of action jazz, but it remains the least popular jazz style.

A decade later, fusion (also known as jazz-rock and crossover music) emerged. With its heavy borrowings of rhythmic, melodic, and instrumental elements from 1960s rock, it quickly became the most popular jazz style since swing. During the 1970s most young jazz musicians seemed to ignore both bebop and action jazz and learned fusion. In addition, many established beboppers switched to the new idiom. It was an understandable choice; fusion groups, on the average, made more money than other jazz groups. Also, a player did not need as strong a grasp of harmony to play most fusion, with its one- and two-chord vamps, as s/he did to play through the sometimes complicated harmonies of bebop tunes.

While these newer styles developed, many of the musicians discussed in the previous chapters continued to play bebop. It seemed destined to become another music of the past, however, since most of the younger players were gravitating to action jazz or fusion, especially the latter. Bebop seemed to be primarily the idiom of players born between 1912 (Gil Evans) and 1943 (George Benson).

But some younger players did go into bebop during those years. One of the finest was (and is) tenor saxophonist Peter Christlieb (born in 1945), who has a big, warm tone, great technical command of his instrument, and an ability to build superb extended solos. Tenor saxophonists Ernie Watts (born in 1945) and Greg Herbert (1947–78) also played great bebop during that decade, though they played fusion as well. Alto saxophonist Richie Cole (born in 1948), who studied under Phil Woods, established a successful career playing Woods-inspired bebop. Another alto saxophonist, Mary Fettig (born in 1953), established herself as a valuable member of Stan Kenton's and Woody Herman's bands. Pianists Richie Beirach (born in 1947) and Onaje Allen Gumbs (born in 1949) devloped thick-textured dissonant styles on the border of bebop and "free" jazz. Trumpeter Jon Faddis (born in 1953) built an impressive career by adopting every nuance of Dizzy Gillespie's style as his own. Alto saxophonist Bobby Watson (born in 1953), tenor saxophonist David Schnitter (born in 1948), and pianist James Williams (born in 1951), all first-rate beboppers, established their credentials as members of Art Blakey's Jazz Messengers.

In the 1980s many more young bebop players gained prominence, and several of them also had ties to Art Blakey. First, and most prominent, were the Marsalis brothers, saxophonist Branford (born in 1960) and trumpeter Wynton (born in 1961), who were in Blakey's Jazz Messengers in 1980 and 1982. They soon established an independent existence as the front line of the Wynton Marsalis Quintet (see below). In 1982, trumpeter Terence Blanchard (born in 1962) and tenor saxophonist Donals Harrison (born in 1960) replaced the Marsalises in the Jazz Messengers. By the mid-1980s, they had formed their own quintet and were touring as co-leaders. At about the same time, OTB (Out of the Blue), a Messengers-inspired group, began recording. This worthy group included trumpeter Michael Philip Mossman, also saxophonist Kenny Garrett, tenor saxophonist Ralph Bowen, and pianist Harry Pickens. In the late 1980s the Harper Brothers (co-led by trumpeter Philip Harper and drummer Winard Harper), a similarly styled quintet, made its debut.

The players just cited are but a few of the many excellent young players who have appeared throughout the decade. Trumpeters Wallace Roney, James Morrison, Brian Lynch, Marlon Jordan, and Roy Hargrove; trombonist Delfeayo Marsalis (a younger brother of Branford and Wynton); saxophonists Vincent Herring, Jim Snidero, Chris Potter,

Ralph Moore, Jeff Clayton, Wes Anderson, Rickey Woodard, Joshua Redman, and Ravi and Oran Coltrane (sons of John Coltrane); pianists Kenny Kirkland, Fred Hersch, Makoto Ozone, Renee Rosnes, Billy Childs, Michel Camilo, Gonzalo Rubalcaba, Marcus Roberts, Kenny Drew, Jr., James Williams, Mulgrew Miller, Benny Green, and Geoff Keezer (the last four are former Jazz Messengers); organist Joey DeFrancesco; guitarists Bruce Forman, Kevin Eubanks, Emily Remler, and Mark Whitfield; bassists John Clayton, Charles Fambrough, Christian McBride, and Charnett Moffett; drummers Jeff Watts and Marvin "Smitty" Smith and others have all demonstrated their mastery of the idiom in the international jazz world. Fusion, the dominating jazz style of the 1970s, shared center stage in the 1980s with bebop played by artists young enough to be the grandchildren of Parker, Gillespie, and Blakey. This renewal of the bebop tradition also inspired such seemingly entrenched fusion players as guitarists Pat Metheny, Joe Beck, and Lee Ritenour to record bebop albums.

To a large extent this bumper crop of fine young players results form the burgeoning jazz education movement in the United States. The Berklee School of Music in Boston, where Branford Marsalis, Makoto Ozone, and others studied, is one of the most famous training schools. Many colleges and universities across the country offer extensive training in jazz performance, as well. And often this training centers on the bebop tradition.

Branford and Wynton, sons of New Orleans bebop pianist Ellis Marsalis, acquired solid instrumental and musical backgrounds while teenagers. They received part of their education from their father at the New Orleans Center for the Creative Arts (where their classmates included Terence Blanchard and Donald Harrison). Branford left home to study, first with Alvin Batiste at Southern University, then at Berklee. Wynton went to New York City to study trumpet at Juilliard. But by 1980, they found the training offered by Blakey was more practical for their needs.

In 1981 they recorded a quintet album, *Wynton Marsalis.*[3] The album sold unusually well, and led to performances on national television and a series of albums—under Wynton's name, under Branford's name, in the jazz idiom, and in the European concert idiom. Columbia, their record company, promoted them heavily; Wynton collected a series of Grammy awards in both jazz and "classical" categories; and the brothers rapidly became top international jazz stars. In 1985, Branford began performing with rock singer Sting, while Wynton continued to divide his career between the worlds of jazz and European concert music. Since then each has led highly successful groups, though they reunite occasionally for a concert appearance or record date.

When he was first in Blakey's band, Wynton Marsalis was a disciple of Freddie Hubbard, Lee Morgan, and Clifford Brown (for example, the Brown-style grace notes in *My Funny Valentine*). Perhaps as a member of Blakey's band he deemed it natural to turn to some illustrious past members of the band for stylistic guidance. He had a few traits of his own, too, such as a lighthearted way of playing rapidly tongued, staccato descents (also in *My Funny Valentine*). At age 18 he showed more promise and considerably more technique than most other famous trumpeters whose teenage efforts are preserved. But Valery Ponomarev, a Blakey veteran of three years, outclasses him in the Blakey album *The Jazz Messengers Big Band*. In general, Marsalis at the time was a player with superb technique searching for a musical way in which to apply it.[4]

A few months later Marsalis showed a new musical personality in his first album as leader, *Wynton Marsalis*. The "new" Marsalis was less inclined to the dramatic and had a distinctive tone quality. Here the influence of Miles Davis of the 1960s outweighs that of Hubbard, especially in *Father Time* and *Twilight*. In *Hesitation* he uses a figure from Davis's solo in *The Theme*.[5] Perhaps again, the surrounding musicians influenced him, for he and Branford recorded part of this album with Herbie Hancock, Ron Carter, and Tony Williams—Davis's rhythm section of the 1960s—while the five of them were touring Japan. Wynton also recorded a double album with the same rhythm section, under Hancock's leadership, during the same tour.[6] In these 1981 recordings Marsalis exhibits several distinguishing traits: 1) the staccato tonguings mentioned earlier, 2) an occasional high-pitched "bip" within a lower-pitched phrase, 3) some intriguing asymmetric rhythms in his chromatic phrases, 4) an impeccably clean tonguing of those chromaticisms, 5) a unique way of half-valving a lower-neighbor tone as he slides up a half-step, and 6) a half-valved rise to a high note followed by a descending line.

Hesitation, from Marsalis's 1981 album, is a fascinating piece. It is a quartet piece by Wynton, Branford, Ron Carter, and Tony Williams, and is another *I Got Rhythm* contrafact. The rhythm team is magnificent throughout, laying down a stimulating rhythmic carpet for Wynton's impish theme and for the musical games that follow it. After trading semi-serious choruses twice, the brothers launch into a lighthearted and startling six-chorus dialogue. But they do not trade 4s, 8s, or phrases of any other fixed length. Instead, they carry on a realistic conversation in which each says what he has to say no matter how long it takes, and then the other responds in similar fashion. For example, in the first of these six choruses, they play phrases of eleven (Branford), ten (Wynton), seven (Branford), and four (Wynton) measures. Later, when the boundary between the second and third of these choruses goes by, Wynton is in the middle of a nine-measure phrase, and no player marks this boundary in an explicit way. The phrase structure coupled with the Marsalises' chromatic and harmonically vague melodic lines make this piece challenging for the listener to follow. It is an altogether delightful experience, however, to keep the tune structure in mind and to compare it with the phrase structure built by the two men.[7]

The brothers recorded one final album with Blakey in 1982, *Keystone 3.*[8] In the piece *A la Mode* from that album, another facet of Wynton's style comes out clearly—an uncommonly beautiful, rich tone quality at a low volume level. It is a hint of what can be appreciated fully only when hearing him play in person and without the distortion of a sound system. His early training and practice regimen give him the ability to play his instrument as efficiently as possible. Using small amounts of air he can play long phrases with a fullness of tone that few others have matched and none has surpassed.

Marsalis recorded his album *Hot House Flowers* in 1984; it is largely a collection of standard popular songs.[9] One such standard is *Star Dust,* which contains probably the most effective use of strings in a jazz context up to that time. The superb, fresh string arrangement by Robert Freeman provided a matchless setting for Marsalis's beautiful sound. Marsalis strikes an exquisite balance between theme paraphrase and improvisation. A few years later he expressed regret over having recorded this album before he was prepared musically for the songs it contains.[10] But it is difficult to imagine him or anyone else improving upon this performance of *Star Dust.*

In 1985 he recorded *Black Codes* and *J Mood,* albums that document another change of style[11] There is a new seriousness in his playing; the impish staccato passages, the tweaked high notes followed by the descents, and the overt demonstrations of virtuosity almost disappear. Also gone is much of the Davis-inspired chromaticism, though strong traces of Davis's style remain. Two of the high points are the most basic of blues in B♭—an untitled duet with Ron Carter in *Black Codes* and the piece *Much Later* in *J Mood.* In both pieces he subscribes to Davis's harmonically unspecific approach to the blues, but there is nonetheless much that is uniquely Marsalis's. For example, Davis did not use the startling 2 + 3-against-4 rhythm that Marsalis plays during his chromatic descent in the eighth chorus of *Much Later:*

In this piece he also initiates some marvelous rhythmic interchanges with Marcus Roberts, whose creative comping is exemplary.

The album *Marsalis Standard Time Vol. 1* shows a continuing stylistic evolution.[12] He retains and exploits to the fullest the strong points of his earlier recordings—the lyricism (especially on those gorgeous long notes), the rhythmic ingenuity (masterfully supported by Roberts and Watts), and the motivic developments. He and his rhythm section give to *Caravan* a New Orleans street beat and a classic performance. In the second chorus of *April in Paris* he launches an extraordinary phrase that extends unbroken for thirteen measures. The phrase begins with 40 beats of triplet-eighth notes and an occasional seven-note grouping stretched over two beats. The melodic twists and turns and the unpredictable scattering of accents in his phrasing create a fascinating melodic cascade.

As the title of the album suggests, most of the pieces are popular and jazz standards. Marsalis had drawn from the older repertory in earlier albums, especially in *Hot House Flowers.* But many of his early recordings were of original compositions in the style that the Davis Quintet and other groups played in the 1960s. Increasingly in the late 1980s and early 1990s he has delved into earlier jazz styles and repertories. His further forays into the standards yielded *Standard Time Vol. 3: The Resolution of Romance,* an album done with his father. He tapped the early New Orleans traditions in *The Majesty of the Blues.* In some concert performances he has played the repertory and solos of Louis Armstrong with elegance, taste, and an obvious reverence for the tradition. He has mastered the art of using the plunger and cup mutes, not just the stemless Harmon of 1950s bebop. Yet no matter what the repertory and the stylistic context, Marsalis's phrasing, rhythmic ingenuity, and indescribably beautiful tone quality shine forth.[13]

Marsalis in his early twenties had the technical capacity to play any notes and rhythms he wanted to play; in his thirties, with his breadth of musical knowledge, he can adapt his personal style to fit any general idiom, and to imprint his musical personality upon it. The possibilities he is equipped to explore, the choices he can make, are more varied than they were for any other major trumpeter in jazz history. It will be fascinating to hear his choices in the years to come.

Branford, too, has enormous possibilities available to him. As bimusical as his brother, he also has made impressive recordings of European concert music. Though he seems flippant and casual compared with his sterner, more intense younger brother, he obviously has a deep appreciation for, and understanding of, the jazz tradition as it relates to the saxophones. And like his brother, his command of his chosen instruments, the soprano and tenor saxophones, seems total.

Branford's personal stylistic direction seems less well defined than his brother's. In the early 1980s his chief role model was Wayne Shorter, an appropriate complement to his brother's admiration of Miles Davis. But he developed a chameleon-like ability to alter tone quality and melodic vocabulary. He learned to imitate several player's styles, much as a comic impressionist imitates Jimmy Stewart or George Bush. Thus, in his albums, one hears creditable imitations of Benny Carter, Ben Webster, Sonny Rollins, middle and late John Coltrane, Ornette Coleman, and others. Some of his playing borders on "free" jazz, while other solos, such as those in *Digital Duke*, are in a conservative bebop style.[14] Doubtless in time he will reveal his ultimate direction. However, his recent career move—in the spring of 1992 he became the highly paid musical director for the *Tonight Show*, where he must display bite-sized samplings of many styles, jazz and non-jazz—may make it difficult for him to define a stylistic focus.

The fame enjoyed by the Marsalis brothers[15] spills over onto some of their sidemen. On the earliest Marsalis album were pianist Kenny Kirkland and drummer Jeff Watts. Kirkland, a muscular pianist influenced to some extent by McCoy Tyner, played with Wynton until he and Branford joined Sting's jazz-influenced rock band in 1985. Then he played in Branford's groups, and also formed his own successful quintet in 1991. Watts, a brilliant percussionist in the tradition of Elvin Jones and Tony Williams, played with Wynton until 1988. Most recently he has played in both Branford's and Kirkland's groups. Much of the rhythmic excitement of the Marsalises' music stems from this powerful and inventive drummer. Both Kirkland and Watts appear on the *Tonight Show* with Branford.

Pianist Marcus Roberts (born in 1963), who replaced Kirkland in 1985, is probably the most distinctive player to emerge from the Marsalis groups. His style owes much to Thelonious Monk;[16] that fact in itself guarantees some stylistic uniqueness, since few pianists have adopted Monk as a principal role model. But Roberts's technique is far different from Monk's, and in his first recordings with Marsalis (*J Mood*) he showed that he also learned some spicy chord voicings from Herbie Hancock, and that he had a rich rhythmic imagination of his own. His comping for Marsalis, as distinctive as his solo style, is remarkable, almost telepathic. The rhythmic rapport between the two men on *Live at Blues Alley* and elsewhere is on a level that Coltrane and Elvin Jones enjoyed in the early 1960s, though the musical results are radically different. In 1989, he left Marsalis to form his own group. One of the albums he has made since then is a solo piano album, *Alone with Three Giants,* devoted to pieces by Jelly Roll Morton, Duke Ellington, and Monk. Clearly he is doing his history homework, just as his former leader has done.[17]

Another young pianist of major significance, one not connected with the Marsalis family, is French-born Michel Petrucciani (born in 1962). He recorded his first album as leader at age 17, and four more before immigrating to the United States at age 20. In his lyrical moments his textures, chord voicings, and beautiful touch suggest Bill Evans.[18] But his style is more muscular, more aggressive, and louder. To get his big sound he has made some adjustments, for he stands only three feet tall.[19] He uses pedal extenders, and he often delivers his forearm's full weight to the keyboard, landing on the outside edge of his little finger, which he reinforces with his fourth finger. His disproportionately large hands have no trouble encompassing octave-spanning chords, however. And his fertile musical imagination regularly generates lengthy solos of high-energy bebop.

As of this year (1994), bebop is thriving. J. J. Johnson, Milt Jackson, Hank Jones, Billy Taylor, Ray Brown, and Max Roach continue to create, as they have done since the beginning of the idiom. Their musical children have flourished for years. Many of their musical grandchildren have established their careers, too; others are busy building local reputations and soon will be heard from on recordings issued by major companies. Will they make important changes in the bebop language? We shall see.

Most of the younger players mentioned in this chapter seem less concerned with breaking new musical ground and more concerned with learning and perpetuating the tradition they have adopted. Often their principal role models are obvious; they include J. J. Johnson (Delfeayo Marsalis), Cannonball Adderley (Herring, Clayton, Anderson, and

Snidero), John Coltrane (Moore), Bill Evans (Hersch), Jimmy Smith (DeFrancesco), and George Benson (Whitfield). In other cases, elements of multiple role models appear in their music. Because their musical roots show, some critics have disparaged their supposed lack of originality. Gillespie's comment at the beginning of the chapter expresses this sentiment. But perhaps, after nearly a century of evolution, jazz has reached a level that satisfies the creative needs of young players learning it. They seek freshness in new compositions and new combinations of old melodic and rhythmic ideas, but do not strive for a new vocabulary or a new jazz idiom. Does that mean that their music is inferior to that of their musical role models?

The answer may lie in society's evaluation of J. S. Bach. Bach invented no musical forms or compositional procedures. Working within the musical vocabulary of his youth, and building upon the structures of works he admired, he created some of the most magnificent musical statements in human history. Many viewed him as old-fashioned during his lifetime. Yet today we treasure his works and barely remember the names of his principal role models.

Is there now a J. S. Bach of bebop who will one day obscure society's memory of these we now regard as jazz giants? Probably not, but writers in the next century may honor some of today's new players as giants themselves. In the meantime, we might well listen with open minds to the music of these young players. If it is good, let us enjoy it without worrying about its evolutionary, or non-evolutionary qualities. Surely, as Frank Wess said, we need not worry about the future of this music; these players are living proof that "Bebop lives!"

Notes

Preface

1. And these are just some fine players who were born in 1945 or earlier; a list of younger players, omitted from Chapter 11, also would be lengthy.

Chapter One: The Beginnings

1. Gillespie/Fraser 1979: 142.
2. Gillespie/Fraser 1979: 208.
3. Levin and Wilson 1949: 1.
4. Ellison 1964: 204 (from an article first published in 1959).
5. Giddins 1987: 78.
6. Louis Armstrong uttered this combination of syllables in his scatted chorus of *Hotter Than That* in 1927 (Okeh 8535).
7. Scott DeVeaux argues convincingly that we should not blame the recording ban alone for the scarcity of early bebop recordings; because of wartime shortages, record companies concentrated on manufacturing only the most popular records, and would have ignored most new jazz anyway (DeVeaux 1988).
8. Most bebop wind players use vibrato much less often than swing players; their long, eighth-note-dominated phrases leave little opportunity for vibrato. Even on longer notes they treat vibrato as an option, not as an effect to be turned on automatically.
9. *Ko-Ko*—6 March 1940, RCA Victor 26577; *Body and Soul*—August 1938, Musidisc 30 JA 5177.
10. On rare occasions in the late 1930s Israel Crosby and Walter Page added some eighth-note embellishments to their bass lines.
11. Gunther Schuller (1989: 111 and 855–56) has an excellent description of how Blanton produced those long tones. He also includes a fascinating computer-generated picture of how those long tones look, as played by Ray Brown.
12. Esoteric 1 and 4 (excerpts are included in the Smithsonian collection). The record producers never paid the musicians, according to Gillespie/Fraser 1979: 139.
13. May 1941, Esoteric 4 and various reissues. (All examples are at concert pitch.)
14. Young recorded *Oh, Lady Be Good!* with Count Basie—9 October 1936, Vocalion 3459; Parker recorded it with Jay McShann—2 December 1940, Onyx 221.
15. Parker's solo dates from the *Lady Be Good* session cited in n. 14; Smith's solo, with the Eddie Durham band, dates from 11 November 1940, Decca 18126.

16. Recorded with McShann—30 November 1940, Onyx 221. I am grateful to Gary Giddins for pointing out this borrowing (Giddins 1987: 61).

17. Clarke discussed his stylistic evolution in an interview conducted for a National Endowment for the Arts Oral History Project; excerpts from that interview appear in Gitler 1985: 52ff., and on this page. Clarke was not alone in trying out new ideas; air checks from 1939 show that Buddy Rich in the Artie Shaw band sometimes used his bass drum to reinforce the brass figures. But Clarke was in the milieu of the "Young Turks" and Rich was not.

18. Recorded on 5 February 1940, RCA Victor 27204.

19. 12 May 1941, Esoteric 1 and other LP reissues; also titled *Charlie's Choice.*

Chapter Two: Early Classics

1. See Gitler 1985: 118ff.; Gillespie/Fraser 1979: 202ff.

2. 23 December 1943, Signature 9001.

3. 13 April 1944, De Luxe 2000; reissued in the Smithsonian set of Gillespie's recordings cited in Chapter 1.

4. 4 and 6 January 1944, Hit 8088, 8089, 8090; reissued on Phoenix LP 1.

5. Mistitled *Blue Garden Blues*—22 August 1944, Hit 7108; reissued on Phoenix LP 1.

6. 5 December 1944, De Luxe 2001 and 2002; reissued in Gillespie's Smithsonian set.

7. 16 February 1944, Apollo 751; reissued in Gillespie's Smithsonian set. Gillespie says that he wrote the piece during the recording session (Gillespie/Fraser 1979: 186). He must have been a speed writer to have written out three choruses for the nine wind players, although he could have sketched a lead sheet (melody and chord symbols) in a few minutes.

8. 15 September 1944, various Savoy 78s and LPs; all reissued on Savoy S5J-5500.

9. 9 January 1945, Manor 1042 and 5000; all four reissued in Gillespie's Smithsonian set.

10. Max Roach says that critic Leonard Feather gave the new music its name because of a communication breakdown. He misunderstood Gillespie's title for his new piece *Be-Bop,* thinking it was Gillespie's generic label for the new music. See Patrick 1975: 21.

11. 26 January 1945, Guild 107; reissued in Gillespie's Smithsonian set. Sarah Vaughan sang on the first recording of this piece, when its title was *Interlude*—31 December 1944, Continental 6031 (reissued on the Smithsonian set). It is a pleasant but bland non-Latin song, whose lyrics complain "love is just an interlude."

12. Gillespie's septet version—22 February 1946, RCA Victor 40-0132; Parker's famous septet version—28 March 1946, Dial 1002.

13. 9 and 28 February 1945, both issued on different pressings of Guild 1001. The first, all but unknown until the 1970s, was reissued in Gillespie's Smithsonian set; the second was reissued on Savoy 12020 and Prestige 24030.

14. "Melodic contrafact" is a term coined by James Patrick to represent the melodic equivalent of *contrafactum,* a vocal composition with a new text replacing the original. See Patrick 1975: 3. Another term for the same practice is "silent theme," proposed by Tirro (1967).

15. 13 April 1944, De Luxe 2000.

16. Many of the early bebop themes were simple and based on riffs, particularly those built on the blues.

17. Savoy 573, 597, and 903; reissued on Savoy compact disc 70737.

18. Discographies, including the one in Gillespie/Fraser, list both Gillespie and Thornton as pianists for take 1 and take 2. For more information about this fascinating recording session, see Gillespie/Fraser 1979: 234, 299–300. After Parker's death, all the master takes, alternate takes, and warm-up pieces were issued on Savoy 12079.

19. Coleman Hawkins uses the same motive in his bebop-style closing theme of *Hollywood Stampede*—2 March 1945, Capitol 10036.

20. 28 March 1946, various Dial 78s and LPs; all reissued on Spotlite 101.

21. 22 August 1947, RCA Victor 20-2480.

22. *Manteca*—30 December 1947, RCA Victor 20-3023. The interest in Latin rhythms went beyond Gillespie during the middle and late 1940s; Woody Herman and Stan Kenton recorded Latin-tinged pieces during the same period. Duke Ellington preceded them by several years, with *The Peanut Vendor, Conga Brava, The Flaming Sword,* and *Bakiff.*

23. 24 December 1947, Columbia 38304. Herman had five men in his saxophone section, but the alto saxophonist had no featured part in this composition. Giuffre's famous piece and Gerry Mulligan's *Disc Jockey Jump,* recorded in January 1947 by Gene Krupa's big band (Columbia 37589), both begin with the same melodic figure.

24. Herman, who never claimed to be a great soloist, played that eight-measure solo note for note for 40 years. However, I saw him play a routine dance job in the early 1980s, when he forgot to enter until the third or fourth measure; by then it was too late to begin the fixed solo, so he actually improvised.

25. 11 January 1949, Prestige 7004.

26. 16 May 1949, Capitol 11060; for a transcription of *Intuition,* a partial transcription of *Digression,* and a detailed discussion of both pieces, see Hellhund 1985: 100ff. A similar early exploration of "free" jazz improvisation is *Abstract No. 1* by Shelly Manne with Shorty Rogers and Jimmy Giuffre—10 September 1954; Contemporary 2516.

27. *Move, Jeru, Budo*—21 January 1949; *Boplicity, Israel*—22 April 1949; all on Capitol 762. For more about these and their companion pieces (a third session took place on 13 March 1950), see Hodeir 1956: 117–36.

28. 4 August 1953, Clef 208; reissued on Verve CDs 825 671-2 and 837 141–2 (the latter is a 10-disc CD set containing everything Parker recorded for Mercury, Clef, and Verve, including many alternate and aborted takes).

29. 30 December 1952, Clef 89141; reissues as in the previous footnote. The singsong title refers to Baird, one of Parker's sons. The grace note and dotted line above the music indicates a slight lagging behind the beat of the rhythm section.

Chapter Three: The Parker Style

1. This quote is from an interview recorded on 15 June 1989, A&M 6404.

2. The unaccompanied solos and some other private recordings—most on tenor saxophone—were issued in 1991 on Stash 535. The broadcast performances and an early jam session performance of *Cherokee* were issued in 1974 under Parker's name on Onyx 221; the best of his performances with McShann are on Decca 79236.

3. When he died some major breakthroughs in recording technology were still in the future. Many of his most influential recordings have serious acoustical deficiencies, particularly in the bass and drum parts. The recordings of the early 1950s capture more of what he and his sidemen produced on their instruments, but they fail to convey the impact he had on ear-witnesses. For example, Paul Bley, who played a concert with him once in 1953, said that "accompanying him was a shock because of the volume. He played about three times louder than I had ever heard anyone play." Lyons 1983: 161; see also Miller 1989: 9–10.

4. The discographical bible for studying Parker is Koster & Bakker 1974–76. A newer discography, Saks, et al. 1989, lacks the extensive details contained in Koster & Bakker. Useful annotations about nearly all of Parker's recordings appear in Koch 1988.

5. A half-step is 100 cents in the Western equal-tempered tuning system. My comments on vibrato stem from some pitch and amplitude graphs of Parker's music made using the UCLA Melograph Model C; see Owens 1974a.

6. From *Salt Peanuts,* recorded on 11 May 1945, Guild 1003.

7. For a fuller discussion of Parker's vocabulary of melodic formulas, see Owens 1974b: I, 17–35 and II, 1–10, where I show a larger, differently numbered catalog of formulas.

8. See Porter 1985: 69. That Parker knew Young's music is shown by, among other anecdotes, Lee Konitz's statement that he heard Parker playing Young's *Shoe Shine Boy* solo in a dressing room during the 1950s. See Gitler 1966: 18.

9. See Baker 1985–86: I, 1, 12; the bebop dominant scale is the Mixolydian scale with an extra note connecting scale degrees 8 and ♭7, and the bebop major scale is the major scale with an extra note connecting scale degrees 6 and 5 (descending lines, especially in Parker's music, are more common than ascending lines).

10. The concept of swing, though hard to define, is central to jazz. For an ensemble to swing the beats must be evenly spaced in time, the rhythm section must synchronize its playing of the beats and the subdivisions of the beat, and the soloist must both synchronize with the rhythm section and create a sense of forward momentum using a skillful manipulation of phrasings and articulations.

11. For details of Parker's life, some of them the stuff in which newsstand tabloids revel, see Reisner 1962, Russell 1973, and Giddins 1987.

12. This exciting performance by Parker has been reissued recently on the ten-compact-disc set, *Bird: The Complete Charlie Parker on Verve,* Verve 837 141–2.

13. The title *Au privave* is meaningless French and probably is the result of a linguistic misunderstanding; perhaps it is a scrambled spelling of *Apres-vous* (used by Max Roach on EmArcy 36127).

14. Levin/Wilson 1949: 12. James Patrick reminded us of the original form of this quote in a book review; see *Annual Review of Jazz Studies 4* (New Brunswick: Transaction, 1988), 208–9.

15. The exact date of this 1948 session remained unknown for years because it took place in violation of the year-long recording ban of 1948 instituted by the American Federation of Musicians. See Leiter 1953: 164–68.

16. For a transcription that attempts to capture the elusive subtleties of this solo, see Owens 1974a: 173–74.

17. The tempo of this last chorus actually drops to ♩ = 60, but the change is hardly noticeable to the ear.

18. King Pleasure's version—24 December 1953, Prestige 880; Supersax's version—1973, Capitol 11177.

19. Recently Gary Giddins noticed that Parker based the first four measures of this solo on the first six measures of *A Table in a Corner;* in the fall of 1939 the big bands of Larry Clinton, Jack Teagarden, Artie Shaw, and Jimmy Dorsey each recorded it, though it soon slipped into obscurity.

20. There was no rehearsal, according to Roach. Four of the players knew one another well, though they had not been playing together during the preceding months; Mingus, however, was new to their music (Gillespie/Fraser 1979: 374).

21. The first release of these pieces was on Mingus's label, Debut. Gillespie, commenting on this record, said, "Do you know that Charlie Mingus . . . took advantage of his position on the bandstand, tape recorded that concert, went home, and put out a record? I ain't seen no royalties until recently" (Gillespie/Fraser 1979: 375). These pieces are now available without the overdubbed bass lines, on the 12-CD set, *Charles Mingus: The Complete Debut Recordings* (Debut 12-4402-2).

22. One indication of his continuing presence was the 1988 feature film *Bird* (released by Warner Brothers, produced and directed by Clint Eastwood with a musical score by Lennie Niehaus), based on his life. The soundtrack is unique in the history of films about musicians, for it contains a number of his actual performances from studio and private recordings, but with new rhythm parts dubbed in to replace the poorly recorded original rhythm sections. (The players who did the overdubbing, like Mingus 35 years earlier, had trouble staying synchronized to the recordings. Jazz musicians, contrary to popular myth, do not keep time with clock-like precision. And the subtle adjustments that players make automatically and subconsciously when playing together can be extremely difficult to replicate once the performance ends.)

Chapter Four: Alto Saxophonists

1. Darroch 1984: 10.

2. Gitler 1966: 42. At the time of the encounter Miles Davis was 17 and Stitt was 19.

3. Musicraft 383, and numerous reissues.

4. "Bird was one of my favorite musicians . . . I haven't heard anybody better. Of course he had an influence on my playing! He influenced everybody in jazz today. . . . I may have a few of Bird's clichés, but I can only be myself" (Bittan 1959: 19).

5. *Wail,* or *Fool's Fancy*—Fall 1946, Savoy 9014; the example is from the first 8 measures of Stitt's solo. *Seven Up*—4 September 1946, Savoy 9006; the example is from the bridge of the first chorus.

6. Design 183.

7. *Hot House—Stars of Modern Jazz Concert;* 24 December 1949; IAJRC 20. *Koko*—29 January 1963; Atlantic 1418.

8. Examples include *All God's Chillun Got Rhythm*—December 1949, Birdland 9001—and *Strike Up the Band*—26 January 1950, Prestige 758 (recorded six weeks after Getz recorded the same piece).

9. *The Eternal Triangle,* with Dizzy Gillespie—19 December 1957, Verve 8262; *Au Privave*—Verve 8344.

10. *On Green Dolphin Street*—Dragon 129/130; *So Doggone Good*—Prestige 10074; *12!*—Muse 5006; *Bye, Bye, Blackbird*—Blackhawk 528-1 (the rhythm section players are Hank Jones, George Duvivier, and Grady Tate).

11. An informal concert recording issued under the name Hollywood Jazz Concert on Savoy 9031; reissued on *Black California,* Savoy 2215.

12. Miles Davis once said: "I remember how at times he [Parker] used to turn the rhythm section around when he and I, Max, and Duke Jordan were playing together. . . . it sounded as if the rhythm section was on one and three instead of two and four. Every time that would happen, Max used to scream at Duke not to follow Bird but to stay where he was. Then eventually, it came around as Bird had planned and we were together again" (Hentoff 1955: 14).

13. *Groovin' High* (from a Gene Norman Just Jazz concert)—29 April 1947, Modern 1201; *Bopera* (from a Hollywood Jazz Concert performance)—April 1947, Bop 108; *Indiana* (from a Harry Babasin jam session that also included Charlie Parker)—16 June 1952, Jazz Chronicles 102; *After Hours Bop* (from another Jazz Concert West Coast performance, issued under Dexter Gordon's name)—April 1947, Savoy 2211.

14. *Sunday*—Imperial 9006 and 9205; the 1959 session with Ola Hansen—Peacock 91 (reissued on Impulse 9337/2); *Willow Weep for Me* and *Paris Blues*—18 August 1967, Prestige 7530; *Blues in My Heart*—24 February 1975, Muse 5068.

15. Detro 1988.

16. As in *Yesterdays,* in Hall Overton's quartet (8 February 1955, Savoy 12146) and in *Falling in Love All Over Again* (25 November 1955, Prestige 7018).

17. *Get Happy* and *Strollin' with Pam*—25 November 1955, Prestige 7018; *Together We Wail* (with George Wallington)—20 January 1956, Prestige 7032.

18. Recorded on 12 August 1957, Signal 1204; reissued on Savoy 12138.

19. July 1957, Mode 127 (mistitled *Billie's Bounce.*) This example begins in the ninth measure of Woods's fourth solo chorus and ends in the third measure of the next chorus.

20. 11 September 1957, Epic 3521.

21. *Altology*—29 March 1957, Prestige 7115; *Pairing Off*—15 June 1956, Prestige 7046.

22. June 1990, Concord 4441.

23. Recorded on 22 December 1961 for Jones; Impulse 11, reissued on MCA Impulse CD 5728. In March 1985 Woods rerecorded this piece with Rob McConnell & the Boss Brass, using a revised version of Jones's earlier arrangement; MCA Impulse CD 5982.

24. *Stolen Moments*—14 November 1968, Pathé 340844; *The Day When the World, Chromatic Bananas,* and *And When We Are Young*—5 July 1970, AZ-Rec 123; Frankfurt Festival pieces—Embryo 530 (reissued on Atlantic 90531-1).

25. *Cheek to Cheek*—November 1976, RCA Victor 2-2202; *Changing Partners*—9 November 1977, Gryphon 782-0798; *Phil Woods Quartet Volume One*—26 May 1979, Clean Cuts 702 (in *Along Came Betty,* he imitates Benny Carter's tone quality and articulation habits); *Phil Woods Quartet Live from New York Village Vanguard*—7 October 1982, Palo Alto 8084; *Gratitude*—19 June 1986, Denon 1316.

26. 5 August 1955, Prestige 7034. The title of this blues theme has been misspelled as *Dr. Jackie, Dr. Jackyl,* and *Dr. Jekyll* over the years.

27. Recorded by George Wallington's Quintet—9 September 1955, Progressive 1001. This piece is a melodic contrafact of *I Got Rhythm* (*a* section) and of *Honeysuckle Rose* (*b* section).

28. Recorded by Jimmy Smith's group in March 1960, Blue Note 4269.

29. That is, emphasizing notes outside the diatonic scale; in this case, emphasizing notes outside the F# Dorian scale.

30. *Cancellation* and *'Snuff*—5 August 1964, Blue Note 4179. The album with Ornette Coleman is *New and Old Gospel*—24 March 1967, Blue Note 4262. Examples of his bebop style include *Plight* and *I Hear a Rhapsody*—16 September 1964, Blue Note 4218; and *Old Gospel*—24 March 1967, Blue Note 4262.

31. Liner notes to *New and Old Gospel.*

32. Mansfield and Green 1984: 17. An important reference work on the early part of McLean's life is Spellman 1966.

33. The published details of Adderley's life appear piecemeal in a number of magazine articles, but are compiled and organized in Kernfeld 1981: I, 180–204. Other important references include an early essay by Adderley for *Jazz Review,* reprinted in Williams 1964: 258–63; an essay by Orrin Keepnews (his record producer during the Riverside years) for the liner notes to *The Japanese Concerts* (Milestone 47029), reprinted in Keepnews 1988: 199–206; and a discography published in *Swing Journal,* October 1975, pp. 248–55.

34. Anonymous 1955: 16.

35. 26 July 1955, Savoy 12018.

36. Parker's version—17 January 1951, Verve 8010; Adderley's version—5 June 1960, Riverside 355.

37. From the last chorus of *Them Dirty Blues*—29 March 1960, Riverside 322.

38. Adderley's solos are in *Spontaneous Combustion, Still Talkin' to Ya,* and *Caribbean Cutie*—14 July 1955, Savoy 12018; Gordon's is in *Daddy Plays the Horn*—18 September 1955, Bethlehem 36. The issue of who invented and who copied is unresolvable.

39. *Our Delight,* etc.—EmArcy 36015 (reissued on CD 826 986-2); *Love for Sale,* etc.—9 March 1958, Blue Note 1595; *Two Bass Hit* and *Miles*—April 1958, Columbia 1193; *A Little Taste*—1 July 1958, Riverside 269; *Groovin' High*—28 October 1958, Riverside 286.

40. Recorded on 18 and 20 October 1959, Riverside 311.

41. *Dat Dere,* etc.—early 1960, Riverside 322; *Sack o' Woe,* etc.—16 October 1960, Riverside 344; *Big "P"* and the second *Azule Serape*—November 1960, Pablo Live 2308 238.

42. The 1962 concerts were in January (Riverside 404), March (Alto 722), and August (Riverside 499); *Work Song*—recorded on 5 August 1962, Riverside 499.

43. *All Blues* and *Flamenco Sketches*—early 1959, Columbia 1355 (reissued on Columbia CD 08163). The titles for these two pieces were switched on the liner notes of the LP, but are listed correctly on the LP label and on the CD reissue.

44. The minor-pentatonic scale consists of degrees 1, $^{\flat}3$, 4, 5, and $^{\flat}7$ of the natural minor scale; the blues scale (actually one of *several* scales so labeled by different writers) adds the half-step between 4 and 5 (the "flatted fifth") to this pentatonic scale.

45. *Mercy, Mercy, Mercy*—July 1966, Capitol 2663; *Walk Tall*—1967, Capitol 2822.

46. 1955, GNP 12.

47. Morgan & Mack 1987: 14.

48. Perhaps it is significant that he plays in C rather than the traditional B$^{\flat}$, for some melodic patterns that in B$^{\flat}$ might come to the fingers almost automatically might not feel right in the higher key.

49. There are so many of Bird's children that to do justice to them all would greatly increase the size of this book. A few more of the many alto saxophonists influenced by Parker's style are John Dankworth, Lou Donaldson, Paquito D'Rivera, Med Flory, Ernie Henry, Derek Humble, John Jenkins, Rahsaan Roland Kirk (who

sometimes played Parker-like solos on the stritch, a rare instrument that looks like a straight soprano sax but is closer in range to the alto), Charles McPherson, Charlie Mariano, Curtis Pegler, Norwood "Pony" Poindexter, Vi Redd, Dave Schildkraut, Dick Spencer, Frank Strozier, Ira Sullivan, and Eddie "Cleanhead" Vinson. Some whose styles are close to that of Phil Woods—Richie Cole, Lanny Morgan, Gene Quill, and Sadao Watanabe—might be thought of as "Bird's grandchildren."

50. 9 March 1956, Prestige 7037.

51. *I Can't Get Started*—16 January 1959, United Artist 4036; *Blues for M.F.*—1962, Roulette 52121.

52. *Dance to the Lady*—15 April 1968, Columbia 9689; *Karuna Supreme*—1 November 1975, MPS 22791; *Hard Work*—January 1976, Impulse 9314.

53. For the record, however, Pepper claims to have been unmoved by Getz's playing, which he regarded as "cold." See Pepper and Pepper 1979: 112.

54. The longest was in the pretentious *Art Pepper*—18 May 1950, Capitol 28008. His best solo during the years with Kenton is *Somewhere Over the Rainbow*—8 October 1951, Capitol 6F-15764—performed with an octet made up mostly of Kenton players and led by Shorty Rogers.

55. They are of a quintet led by Shorty Rogers at the Lighthouse—27 December 1951, Xanadu 148—and a quartet led by Pepper—12 February 1952, Xanadu 108 and 117.

56. My measurements of vibratos applied to long notes played by a variety of players yield the following average rates: typical swing-era players (Hawkins, Hodges, and others)—about 5.6 oscillations per second; Lester Young—about 4.5 ops; Art Pepper—about 4.2 ops; Charlie Parker—about 4.9 ops; typical bebop players—4.7 ops. These differences might seem unimportant, but they are aurally perceptible, just as the difference between tempos of ♩ = 250 and ♩ = 290 is clear to the ear. (These are the tempos of Pepper's and Parker's vibratos expressed in another way.)

57. See Pepper and Pepper 1979: 477–507—compiled by Todd Selbert.

58. 26 November 1956, Contemporary 7630; reissued on Contemporary CD 1577, with additional pieces and alternate takes.

59. 19 January 1957, Contemporary 7532.

60. Contemporary 7568.

61. Significantly, Med Flory, the lead alto saxophonist in Supersax, was the baritone sax player on these Pepper-Paich sessions.

62. 29 February 1960, Contemporary 7573.

63. He also builds his solo in *Smack Up*—October 1960, Contemporary 7602–out of a single motive.

64. 13 August 1981, Galaxy 5145.

65. Pepper exerted an influence of his own, especially on Herb Geller, Joe Maini, and Lennie Niehaus, all based in Los Angeles.

66. See Gitler 1966: 258 and Kirchner 1978: 10.

67. *Anthropology*—4 September 1947, Columbia 38224; *Yardbird Suite* (the source of the disjunct musical example)—17 December 1947, Columbia 39122.

68. From *Marshmallow* (based on the chords of *Cherokee*—28 June 1949, New Jazz 807.

69. 7 April 1950, New Jazz 834 and 827; reissues on Prestige.

70. World Pacific 20142.

71. This figure is one that pianist Lennie Tristano and alto saxophonist Paul Desmond enjoyed playing as well.

72. June 1955, Atlantic 1217.

73. 29 August 1961, Verve 8399; reissued on Verve CD 821 553–2, with three added pieces.

74. 25 September 1967, Milestone 9013.

75. 25 May 1982, Owl 28.

76. On occasion his love of the sequence extends beyond the limits of good judgment; some of them go on too long in *The Song Is You*—March 1954, Columbia 566.

77. A good example is his ingenious solo in *Pennies from Heaven*—22 February 1963, Columbia 2CL-26.

78. 14 December 1953, Fantasy 3-13.

79. All recorded in March 1954, Columbia 566. Desmond, unlike Konitz, was inspired by the blues structure; his solos in *St. Louis Blues* and *Bru's Blues*—May 1957, Columbia 1034—are also excellent.

80. Columbia 1397.

81. Early 1977, Horizon A&M 726.

Chapter Five: Tenor Saxophonists

1. 11 October 1939, Bluebird 10253. See Schuller 1989: 441ff. for an excellent transcription and analysis of this performance.

2. For an extended and perceptive discussion of Hawkins's music, see Schuller 1989: 426ff.; the most complete biography is Chilton 1990.

3. 9 October 1936, Vocalion 3459. For an excellent analysis of this solo, see Schuller 1989: 230ff.

4. *Lester Leaps In*—5 September 1939, Vocalion 5118. Two fine books on Young and his music are Porter 1985 and Porter 1991. For more on some of the melodic figures Hawkins and Young contributed to early bebop, see pp. 5–6, 9.

5. *I Got Rhythm*—9 June 1945, Jazz Star 47101; *I Found a New Baby*—21 August 1946, Savoy 627.

6. 22 April 1947, RCA Victor 20-2504.

7. His first recording on the soprano saxophone predates that of John Coltrane by more than a year.

8. The best reference works on Gordon are the marvelous discography by Sjøgren (1986) and the recent biography by Britt (1989).

9. December 1947, Savoy 2211.

10. The best samples of his blues playing in the mid-1940s are the two takes each of *Blow, Mr. Dexter* (30 October 1945) and *Long, Tall Dexter,* a title later shortened to *LTD* (29 January 1946). All four appear on the LP reissue Savoy 2211.

11. 18 September 1955, Bethlehem 36.

12. *Lovely Lisa*—13 October 1960, Jazzland 29; *You've Changed*—6 May 1961, Blue Note 4077.

13. 9 May 1961, Blue Note 4083.

14. The jazz world owes recording engineer Rudy Van Gelder a great debt of gratitude for recording hundreds of bebop albums. At first a full-time optometrist, Van Gelder had a recording studio, complete with a small Steinway grand piano, set up in his living room in Hackensack, New Jersey. In 1953 he became the principal recording engineer for Blue Note Records. Soon he was engineering for Prestige and Savoy as well; and in the 1960s he added Impulse to his list of employers. In 1959 Van Gelder abandoned optometry, moved to Englewood Cliffs, New Jersey, and built a new studio

(Cuscuna 1984). His skill at getting a proper mix of instruments directly onto the master tape (long before multiple-channel recording existed) was exemplary, and his clean, crisp, well-balanced drum-kit sounds were especially noteworthy. Perhaps his most distinctive aural signature was a tight, boxy piano sound.

15. 27 August 1962, Blue Note 4112.

16. 23 May 1963, Blue Note 4146.

17. Recorded with the Ellington band—4 May 1940, RCA Victor 26610.

18. "Jaw trill" is perhaps a better term than "lip trill," for, according to Trent Kynaston, the effect comes from combining an unusual fingering with an "exaggerated, rapid jaw movement" (Kynaston 1992: 59). Saxophonist Bill Green has told me that several factors must come together to produce this effect: a false (i.e., unusual) fingering, an unusual setting of the mouth, neck, and throat, and the aforementioned exaggerated jaw movement. Alto saxophonists produce the wobble on D♭ (again, fingered as B♭ on that instrument.).

19. 2 June 1964, Blue Note 4176.

20. *Sonnymoon for Two*—20 July 1967, Black Lion 108; *Tenor Madness*—23 August 1975, Steeplechase 1050.

21. The first half of this phrase has been in jazz since at least the 1920s.

22. 23 March 1979, at the Keystone Korner in San Francisco, Blue Note 85112.

23. The 11-man ensemble—June 1977, Columbia 34989; *Body and Soul*—May 1978, Columbia 35608; *Come Rain or Come Shine*—13 May 1978, Blue Note 85112; TV show—1979 (part of the *Jazz at the Maintenance Shop* series); *Skylark*—16 March 1982, Elektra/Musician 60126.

24. Released in 1987, Columbia 44029.

25. 5 December 1944, Deluxe 2001.

26. *Goodbye*—March 1974, Prestige 10093.

27. For example, the album *Boss Tenors*—August 1961, Verve 8426.

28. These are the words of pianist Jimmy Bunn, who played piano for some of these tenor battles, from a telephone conversation I held with him on 22 August 1989.

29. *The Chase*—12 June 1947, Dial 1017 (which contains the same introduction as Charlie Parker's *Klact-oveeseds-tene*—4 November 1947, Dial 1040); *The Hunt* (listed under the name Hollywood Jazz Concert)—6 July 1947, Bop 101, 102; *The Steeplechase* (recorded for Gene Norman)—2 February 1952, Decca 7025.

30. Gray's indebtedness extended to using the first nine measures of Young's classic 1936 solo on *Oh, Lady Be Good* at the opening of his solo in *Backbreaker* (April 1947, Savoy 9031), a melodic contrafact of *Oh, Lady Be Good.*

31. 11 November 1949, Prestige 115.

32. In *I Got Rhythm* in a JATP concert—22 April 1946, Mercury 35014.

33. Another obvious bow to Parker occurs when he begins his solo on *Scrapple from the Apple* (27 August 1950, Prestige 759 and 760) with the first four measures of Parker's solo on the same piece.

34. Edwards was a sideman on several sessions for Howard McGhee; the pieces are on Modern Music, Aladdin, and Dial 78s (the latter are reissued on Spotlite 131). His first session as leader is on the obscure Rex label, reissued on Onyx 1215.

35. August 1959, Pacific Jazz 6.

36. In his solo on the alternate take of *Grab Your Axe, Max* (14 December 1945, Savoy 1105), he quotes Young's first phrase in the Basie recording of *Jive at Five.*

37. 31 July 1946, Savoy 903 and 954. *Opus de Bop* is built on the chords of *I Got Rhythm; Running Water* is built on the chords of *Running Wild.* The wider the gap—or "lay"—between the tips of the mouthpiece and reed, the bigger the sound; the lay of the

mouthpiece and the material from which it is made (metal or rubber) play vital roles in determining the basic tone quality of a saxophone.

38. 30 December 1948, Capitol 57616.

39. 21 June 1949, New Jazz (Prestige) 811. In jazz, especially prior to the 1970s, the most common keys used for the major mode are G, C, F, $B^{\flat}$, $E^{\flat}$, $A^{\flat}$, and $D^{\flat}$. Some pieces touch briefly on the remaining keys, especially in the bridge sections, but many players would want some extra practice before performing a blues in $F^{\#}$, B, E, A, or D.

40. 6 January 1950, Prestige 740.

41. *Mosquito Knees*—28 October 1951, Roost 103; *Lover, Come Back to Me*—12 December 1952, Clef/Mercury 89042.

42. For a collector of Getz recordings the discography by Astrup (1984) is indispensable.

43. In an interview in 1967 Getz stated that *Focus* was his favorite album (Jones 1987: 47).

44. Among Sims's best recordings are *Zoot at Ease*—1973, Famous Door 2000—and a duet album with guitarist John "Bucky" Pizzarelli entitled *Summun*—August 1976, Ahead 33752. In the 1950s Sims doubled on alto saxophone occasionally and achieved a sound remarkably similar to that of Art Pepper. Some of Cohn's best recordings include *Al Cohn's America*—6 December 1976, Xanadu 138—a duet album with pianist Jimmy Rowles entitled *Heavy Love*—15 March 1977, Xanadu 145—and *Nonpareil*—April 1981, Concord Jazz 155.

45. Recorded with Lee Konitz—14 June 1955, Atlantic 1217. The $A^{\flat}$ between those notes is the highest note before entering the altissimo register on the alto saxophone.

46. 29 June 1954, Prestige 187. *Airegin* is "Nigeria" written backwards; *Oleo* is a melodic contrafact of *I Got Rhythm; Doxy* is a contrafact of *How Come You Do Me Like You Do.*

47. *Dig,* with Miles Davis—5 October 1951, Prestige 777; *No Line,* with Davis—16 March 1956, Prestige 7044; *Tenor Madness*—24 May 1956, Prestige 7047.

48. *Shadrack*—17 December 1951, Prestige 7029; *Silk 'n' Satin*—18 August 1954, Prestige 186; *There's No Business like Show Business*—2 December 1955, Prestige 7020; *How Are Things in Glocca Morra?*—16 December 1956, Blue Note 1542; *I'm an Old Cowhand* and *Wagon Wheels*—7 March 1957, Contemporary 3530; *Toot, Toot, Tootsie* and *The Last Time I Saw Paris*—June 1957, Riverside 241.

49. *Reflections*—14 April 1957, Blue Note 1558; *Blue Seven*—22 June 1956, Prestige 7079; *Ee-ah*—7 December 1956, Prestige 7126; *Wonderful! Wonderful!* and *Blues for Philly Joe*—22 September 1957, Blue Note 4001; *Sonnymoon for Two*—3 November 1957, Blue Note 1581.

50. On Prestige 7044 and 7047 (see n. 47).

51. There appears to be one moment of confusion near the end of the performance, when neither Rollins nor pianist Tommy Flanagan play; bassist Doug Watkins, perhaps thinking at first that the moment of the coda has arrived, plays four extra measures in the first of two time-marking choruses.

52. *St. Thomas*—1959, Dragon 73; *Brownskin Girl*—1962, RCA Victor 2572; *Hold 'em, Joe*—8 July 1965, Impulse 191; *The Everywhere Calypso*—July 1972, Milestone 9042; *Don't Stop the Carnival*—April 1978, Milestone 55005; *Little Lu*—May 1980, Milestone 9098; *Duke of Iron*—September 1987, Milestone 9155-2. Rollins's interest in calypso music stems from his mother, who was born in the Virgin Islands.

53. *Body and Soul*—10 July 1958, Metrojazz 1002; *Skylark*—July 1972 Milestone 9042. The precedent for unaccompanied saxophone solos is Coleman Hawkins's 1948 recording of *Picasso* (Jazz Scene unnumbered; Clef 674).

54. *The Most Beautiful Girl in the World* (with Max Roach)—12 October 1956, EmArcy 36108; *Stay as Sweet as You Are*—4 March 1959, Dragon 73.

55. The fastest recording I know by Rollins is his performance of *B-Quick*—7 December 1956, Prestige 7126—played almost effortlessly at the astonishing tempo of ♩ = 425. For the sake of comparison, Parker's famous *Koko* of 1945 (both pieces share the harmonic structure of *Cherokee*) is considerably slower: ♩ = 300.

56. He achieves his timbre without resorting to loud playing. Once, during an engagement at Concerts by the Sea in Redondo Beach, California, he left the stage in the middle of a solo and moved through the club while playing. As he stood next to me, playing with the bell of his instrument pointed at my head, I was surprised at how quietly he was playing, all the while generating that famous, intense sound.

57. *Paradox*—2 December 1955, Prestige 7020; *Way Out West,* recorded in Los Angeles—7 March 1957, Contemporary 3530; Village Vanguard albums—Blue Note 1581 and 61015.

58. 11 February 1958, Riverside 258.

59. The first phrase of the last main section resembles a phrase in Stravinsky's *Petroushka,* and may be the inspiration for John Coltrane's *Cosmos* of 1965.

60. March 1959, Dragon 73.

61. *The Night Has a Thousand Eyes*—5 April 1962, RCA Victor 2572; the album with Don Cherry—July 1962, RCA Victor 2612. Rollins had explored action jazz as early as 7 December 1956 (Prestige 7126), before Ornette Coleman and others had recorded.

62. *Playin' in the Yard*—July 1972, Milestone 9042; *Lucille* (his wife's name)—September 1975, Milestone 9064; *Camel*—April 1978, Milestone 5005. For discographical data on Rollins, see Sjøgren 1983.

63. Kernfeld 1988: I, 235; Kernfeld's article on Coltrane gives an excellent overview of the man and his musical idiom. For a meticulously assembled discography, see Wild 1979.

64. In July 1946, Coltrane and some fellow amateur musicians who were serving in the U.S. Navy in Hawaii made some private recordings of pieces that Parker had recorded during 1945–46. *Koko,* from this series, was used on the American Public Radio's *The Miles Davis Radio Project* in 1990, and *Hot House* appears on a two-disc Coltrane anthology issued in 1993 (Rhino/Atlantic R2 71255). These early performances indicate that, at age 20, Coltrane was an unpolished bebopper who had found virtually none of his mature-period tone quality, phrasing, or melodic formulas. *We Love to Boogie*—24 February 1951, Dee Gee 3600 (reissued on Savoy 2209); *Groovin' High*—20 January 1951, on private tapes only; *Jumpin' with Symphony Sid*—17 March 1951, on private tapes only; *Birks' Works*—3 February 1951, Oberon 5100.

65. 6 January; Oberon 5100.

66. *Ah-Leu-Cha*—27 October 1955, Columbia 949; *Miles' Theme*—16 November 1955, Prestige 7014.

67. Dexter Gordon, Miles Davis, and others used the same quotation in the early 1950s. I thank Jimmy Rowles for remembering this 1940 Jimmy van Heusen tune.

68. 11 May 1956, Prestige 7166.

69. Both recorded on 23 May 1958, Prestige 7316. Compared with Coltrane's dazzling technical display, the eighth-note-dominated solos by Donald Byrd and Red Garland on these recordings seem pallid, even though both men play well.

70. Transcribed from the interview recorded in August 1966. Although Kofsky (1970: 221–43) transcribed the majority of the conversation, one can appreciate the

quiet, modest, and deliberately paced conversation of Coltrane much better by listening to his voice. Pacifica Program Service (5316 Venice Boulevard, Los Angeles, CA 90019) sells copies of the tape.

71. The diminished scale, which players also know by several other names, is an eight-note scale consisting of half-steps alternating with whole steps. The name refers to the fact that notes 1-3-5-7 and notes 2-4-6-8 form diminished-seventh chords. Coltrane probably derived his figure 8 from scale pattern #448 on p. 58 of Nicolas Slonimsky's *Thesaurus of Scales and Melodic Patterns* (1947), a reference work that he studied intensely in the 1950s.

72. See Kernfeld 1981: I, 38–58, and II, 214–34, for an excellent discussion of the formulaic side of Coltrane's style in the late 1950s. Lewis Porter's dissertation (1983), although focused primarily on the later music, also has much important information on Coltrane's music of the 1950s (pp. 56–113). A short discussion of Coltrane's patterns is in Sickler 1986: 8–11; a lengthy table of patterns, all transposed to C, is in Baker 1978: 63–87.

73. Pieces from the May session—scattered among Prestige 7166, 7200, and 7129. Pieces from the October session—scattered among Prestige 7129, 7200, 7150, 7166, and 7094.

74. This phrase was coined by Ira Gitler, for his 1958 notes for Prestige 7142, to describe the relentless flurries of notes that Coltrane often played in his solos during the late 1950s.

75. On Prestige 7188 and 7378.

76. In recent years Andrew White (4830 South Dakota Avenue, N.E., Washington, DC 20017), who has transcriptions of over 400 solos for sale, has lightened the task of learning Coltrane's solos. Smaller collections of published solos include those by Sawai, Watanabe, Baker, and Sickler.

77. Lewis Porter (1983: 65) points out an early example of motivic playing in the first portion of Coltrane's solo in *Traneing In* (23 August 1957, Prestige 7123).

78. April 1958, Columbia 1193; reissued on Columbia CD set C4K 45000.

79. Kernfeld (1981: I, 150–54), in a perceptive discussion of this piece, points out the tonal ambiguity created when the different players emphasized different notes of the set of pitches included in the F-major scale.

80. The layout of the annotated transcription of this solo in Kernfeld 1981 (II, 235–38) is helpful in seeing motivic developments and relationships.

81. Carr 1982: 109.

82. Ozone 18; Coltrane's solo is shown on the video documentary *The Coltrane Legacy*.

83. *Giant Steps* and the better take of *Countdown*—Atlantic 1311; *Countdown*, alternate take (and an inferior version of *Giant Steps* from an earlier session)—Atlantic 1668.

84. *Fifth House*—2 December 1959, Atlantic 1354; *But Not for Me*—26 October 1960, Atlantic 1361; *Body and Soul*—24 October 1960, Atlantic 1419. Coltrane's reharmonization of *Body and Soul* has itself become a standard way of playing the piece. Bebop musicians, when playing it together for the first time, either discuss ahead of time whether to play the "Coltrane changes" or else watch and listen *very* carefully during the first chorus! The III-V-I progression appears in *Sambo*, a composition by Shorty Rogers, recorded by the Kenton band—19 September 1951, Capitol 20244. But Coltrane probably came upon the progression by working with the chords shown with scale #182 in Slonimsky's *Thesaurus* (1947: 27). And the harmonization of the second eight measures

of *Giant Steps*—a series of ii-V-I chords a major third apart—probably derives from a harmonization Slonimsky used for scale #184.

85. Ballad collections—1961–62, Impulse 32, and 7 March 1963, Impulse 40 (with singer Johnny Hartman); Ellington collaboration—26 September 1962, Impulse 30.

86. *Equinox*—26 October 1960, Atlantic 1419; first recording of *Impressions*—1 November 1961, Impulse 9361-2.

87. Their initial recording (21 October 1960, Atlantic 1361) pales in comparison with the intensely energetic version of 26 October 1963 (Pablo 2620 101) and others on bootleg labels.

88. *Main Spring*—23 October 1957, Blue Note 1580; *Terry's Tune*—February 1958, Riverside 274; *Blues March* (with Blue Mitchell)—July 1958, Riverside 273.

89. *Rhythm-a-ning*—7 August 1958, Riverside 262; *Blue 'n' Boogie* and *Cariba*—25 June 1962, Riverside 434 (reissued on Riverside CD 106-2); *Blues Up and Down*—23 September 1978, Columbia 35978.

90. November 1970, CTI 6005.

91. All three pieces were recorded around 1959–60, Time 52086.

92. *Exodus*—1961, Vee Jay 378; *Freedom Jazz Dance*—30 August 1965, Atlantic 1448; *Eddie Who?*—27 February 1986, Impulse 33104.

Chapter Six: Trumpeters

1. The best Gillespie discography is Koster & Sellers 1985, 1988.

2. The derivation was secondhand, however, for according to Gillespie, he learned several of Eldridge's solos from Charlie Shavers, who had learned them from Eldridge's recordings. Gillespie and Shavers were section mates in Frankie Fairfax's band in 1935 (Gillespie/Fraser 1979: 54).

3. In Chapter 3, I outlined Charlie Parker's musical idiom, then used that outline as the basis for discussing other saxophonists. I find it convenient to continue this comparison while discussing other groups of instrumentalists, even though this approach may imply that most bebop players imitated Parker. Some players copied Parker directly, others copied players who copied Parker, and a few, such as Gillespie, copied the swing-era players that Parker copied.

4. Recorded with Lucius "Lucky" Millinder's big band—29 July 1942, Brunswick 03406.

5. Recorded on 15 February 1943 in a Chicago hotel room; Stash 260 (issued in 1986) and 535.

6. Gillespie preferred to call this chord a minor-sixth with the sixth in the bass—$B^{\flat}\text{MI}^{(\text{ADD }6)}/G$. See Gillespie/Fraser 1979: 135, 186.

7. In this recording the ii-V chords are in the solo line only, not in the accompaniment; but in subsequent recordings by Gillespie (and most other jazz musicians) the accompanists follow Gillespie's colorful harmonies.

8. 29 December 1945, Spotlite 123.

9. In his later years his technique declined, and these quick-note phrases gave the impression of "marbles loose on stairs," as trumpeter Bobby Shew described them to me.

10. 9 July 1946, Musicraft 404.

11. 6 February 1946, Dial 1004.

12. Some bootleg recordings of the band contain extended solos, but the recording quality is poor.

13. 22 August 1947, RCA Victor 20-2603.

14. The audience for big-band bebop was disappointingly small in many places in the U.S. Once in 1950 Gillespie's big band played a concert in Little Rock, Arkansas, for an audience of only two dozen people sitting in a concert hall with 5000 seats. Thomas 1975: 50.

15. They include (to list only those with the clearest connection with Gillespie) Benny Bailey, Guido Basso, Clifford Brown, Pete and Conte Candoli, Miles Davis, Kenny Dorham, Jon Faddis, Ray Gonzalez, Benny Harris, Russell Jacquet, Howard McGhee, Blue Mitchell, Fats Navarro, Red Rodney, and Clark Terry.

16. 29 July 1946, Dial 201. Parker was arrested that evening and subsequently committed to Camarillo State Hospital (Russell 1973: 218–41).

17. *High Wind in Hollywood*—18 October 1946, Dial 1012; *Hot and Mellow* and *Messin' with Fire*—1947, Swingmaster 14.

18. May 1961, Contemporary 3588.

19. The piece is *Fly Right,* which is actually the first recording of Thelonious Monk's *Epistrophy,* recorded by Cootie Williams's big band—1 April 1942, Columbia 3L 38. Despite the early recording date, the recording was unknown until its belated release in the 1960s. For more on Guy, including a transcription of the triplet figure, see Schuller 1989: 403–4.

20. *Everything's Cool* by the Be-bop Boys—6 September 1946, Savoy 586.

21. *Let's Cool One*—30 May 1952, Blue Note 1602; *Hippy*—6 February 1955, Blue Note 5062; *Soft Winds*—11 November 1955, Blue Note 1507; *Ezz-Thetic*—12 October 1956, EmArcy 36098; *Blues Elegante*—4 April 1956, ABC Paramount 122.

22. For the fascinating story of Rodney's wild life, see Giddins 1981: 228ff. (a reprint of an article from January 1975). For an excellent sampling of his mature style, listen to the album *No Turn on Red*—August 1986, Denon 73149.

23. A good example of his *Perdido* performances dates from 8 September 1959, Columbia 1400.

24. Two fine examples are *Moanin'* with Jones—26 May 1959, Mercury 71489; and *Let My People Be* with Mulligan—December 1960, Verve 8396.

25. Recorded in 1964 by the quintet Terry co-led with valve trombonist Bob Brookmeyer–Mainstream 320.

26. *Melody Maker,* 4 September 1954. Jack Chambers suggests that Eckstine may have exaggerated Davis's inability. Chambers 1983: 20.

27. According to Ian Carr, Davis's tone quality stems in part from his career-long use of a Heim mouthpiece, the unusually deep cup of which helps "in the production of a full sound" (Carr 1982: 6).

28. *Ornithology*—28 March 1946, Dial 208; *Cheryl* and *Buzzy*—8 May 1947, Savoy 952 and MG 10001.

29. All takes are available on Savoy S5J-5500, released under Parker's name.

30. *Bongo Bop*—28 October 1947, Dial 1024; *Out of Nowhere*—4 November 1947, Spotlite LP 105; *Drifting on a Reed*—17 December 1947, Dial 1043; *Ornithology*—Savoy 12179; *Big Foot*—Savoy 12186.

31. This example is from the last 8 measures of his solo in *Boplicity*—22 April 1949, Capitol 60011.

32. 24 December 1954, Prestige 7150.

33. 5 October 1951, Prestige 766.

34. Prestige 777 and 846. At the end of *Bluing* (it is spelled with an "e" on the record label and without an "e" in the liner notes), drummer Art Blakey continues playing after the rest of the band stops. Davis then complains that they will have to record the piece again because of Blakey's mistake; but apparently the producer and Davis decided that the solos were good enough to warrant releasing this flawed performance.

35. 19 May 1953, Prestige 902. Davis repeated the error three years later with a different band—26 October 1956, Prestige 7094.

36. Recordings by Art Blakey, Kenny Clarke, Herbie Mann, Oscar Peterson, and others used the erroneous Davis bridge; see Berger et al. 1982: II, 338ff.

37. Unfortunately, the anonymous compilers of the *Real Book,* a well-known bootleg fake book, used the Davis version of Monk's tune, thereby contributing enormously to the spread of the faulty version.

38. 16 November 1955, Prestige 7014.

39. *I Didn't* is based on Davis's faulty reharmonization of Monk's *Well You Needn't* (Prestige 7007).

40. For this and the next example: 7 June 1955, Prestige 7007.

41. 16 November 1955, Prestige 7014.

42. A fine discussion of Davis's ballad art is in Brofsky 1983.

43. *Airegin*—October 1956, Prestige 7094; *Dr. Jackle*—April 1958, Columbia 1193; *Walkin'*—October 1960, Dragon 129/130.

44. This example is from measures 4–6 of his last chorus in *Blues for Pablo*—23 May 1957, Columbia 1041.

45. See also Chapters 4, 5, and 7 for more about these pieces.

46. Barry Kernfeld points out that bassist Paul Chambers uses D Dorian, but E♭ harmonic minor (which contains D rather than D♭), in playing the theme (Kernfeld 1981: 146).

47. 2 April 1959, Ozone 18. During the same show Davis also performed three of the *Miles Ahead* compositions, with a large ensemble arranged for and led by Gil Evans.

48. 13 October 1960, Dragon 129/130.

49. An excellent analysis of his musical vocabulary for the blues in F—*Walkin'* is such a blues—appears in Kernfeld 1981: I, 105–27.

50. *Teo*—21 March 1961, Columbia 8456; *Mood*—22 January 1965, Columbia 9150; *Masqualero*—17 May 1967, Columbia 9532; *Spanish Key*—20 August 1969, Columbia 30121; *Fat Time*—ca. March 1981, Columbia 36790.

51. The example is from the sixth chorus of *Footprints*—25 October 1966, Columbia 9401.

52. *Kix*—27 June 1981, Columbia 38005. *Mr. Pastorius*—ca. 1989, Warner Bros. 9 25873-2. *Dingo*—1990, Warner Bros. 9 26438-2. In the feature film *Dingo,* Davis played a pivotal on-camera role; he wore a wig and looked much as he did in the early 1970s. In July 1991, at Montreaux, he and a large ensemble performed music from his famous collaborations with Gil Evans in the 1950s; the spirit and some sound of the original performances emerged, but the difficult music was under-rehearsed and Davis's playing was faltering.

53. Collections of transcribed solos by Davis include Nakagawa 1975, Baker 1976b, Julien 1984, and Brown, J. R. 1987.

54. He played a one-chorus solo in *I Come from Jamaica* and a half-chorus solo in *Ida Red* (Okeh 6900 and 6875; both are reissued on Columbia 32284).

55. 12 July 1954, Pacific Jazz 19.

56. During a question-and-answer session at El Camino College on 24 January

1988, Gillespie stated that Brown played piano very well, in a style similar to Bud Powell's.

57. 12 July 1954, Pacific Jazz 19.

58. Elektra Musician 60026, recorded in a Chicago night club.

59. In a meticulous examination of *I Can Dream, Can't I?*, take 1 (15 October 1953, Vogue 5177), Milton Stewart distinguishes several different levels of tonguing, "from a violent slap . . . to a legato caress . . ." (Stewart 1974–75: 147).

60. 7 November 1955, Columbia 35965.

61. January 1955, EmArcy 36005.

62. The first example is from *Carvin' the Rock*, take 14, and the second is from *Cookin'*, take 7; from a Lou Donaldson session—9 June 1953, Blue Note 84428.

63. *Willow Weep for Me*, from the strings album—20 January 1955.

64. Weir (1986) is an important discographical resource for studying Clifford Brown's art. Several collections of transcribed solos exist, including those by Slone (1982) and Yoshida (n.d.).

65. Among those who based part of their playing styles on Brown's are Tom Brown, Dave Burns, Donald Byrd, Ian Carr, Art Farmer, Bill Hardman, Tom Harrell, Freddie Hubbard, Carmell Jones, Booker Little, Blue Mitchell, Lee Morgan, Sam Noto, Valery Ponomarev, Woody Shaw, Idrees Sulieman, and Charles Tolliver.

66. From a Hank Mobley session—5 November 1956, Savoy 12091.

67. *The Sidewinder* and *Gary's Nightmare*—21 December 1963, Blue Note 4157; *Like Someone in Love*, with Blakey—7 August 1960, Blue Note 4245; *Take Twelve*—24 January 1962, Jazzland 80; *The Sixth Sense*—10 November 1967, Blue Note 4335. For discographical information, see Wernboe 1986.

68. The pieces from this session are scattered among Prestige 7130, 7181, and 7209.

69. *I'm a Fool to Want You*—21 September 1961, Blue Note 4101; *Free Form*—11 December 1961, Blue Note 4118.

70. *Open House*, with Jimmy Smith—22 March 1960, Blue Note 4269.

71. *Peace* and *Sister Sadie*—10 August 1959, Blue Note 4017; *Nica's Dream*—9 July 1960, Blue Note 4042; *Filthy McNasty*—May 1961, Blue Note 4076; *Silver's Serenade*—April 1963, Blue Note 4131.

72. *Satin Soul*—ca. 1975, RCA Victor 1109; *Mississippi Jump*—1977, Impulse 9328.

73. *Red Clay*—January 1970, CTI 6001.

74. *Stolen Moments*, recorded with Oliver Nelson–23 February 1961, Impulse 5.

75. *Blues for Duane*—9 December 1969, MPS 15267.

76. See Coleman, *Free Jazz*—21 December 1960, Atlantic 1364; Coltrane, *Ascension*—28 June 1965, Impulse 95; Mimaroglu, *Sing Me a Song of Songmy*—21 January 1971, Atlantic 1576; Hubbard, *Windjammer*—1976, Columbia 34166.

Chapter Seven: Pianists

1. Monk, more than any other musician in jazz history, must be seen to be appreciated fully. Only after seeing his sweating, foot-stomping intensity at the keyboard can one understand his involuntary humming. Only after watching his hands and arms can one understand why his melodic lines sound as they do. Only after watching his slow dervish-like spinning and other personal idiosyncrasies can one understand the

melodic, rhythmic, and harmonic eccentricities of his compositions and improvisations. Fortunately, there is video footage readily available: *Music in Monk Time* (1983, Video and the Arts) and *Straight No Chaser* (1988, Warner Bros.). The second of these is perhaps the best documentary film ever made on a jazz musician. Both are essential parts of the Monk legacy.

2. Monk may have gotten this figure from Duke Ellington; see Schuller 1989: 131.

3. Recorded with Miles Davis on 24 December 1954, Prestige 196.

4. Hodeir (1962: 174) calls this "the first formally perfect solo in the history of jazz." For additional discussion of this solo, see Blake 1982: 24ff.

5. 24 December 1954, Prestige 200.

6. Surviving private recordings made at Minton's in 1941 show him playing much in the style of Teddy Wilson and other stride players, but with a heavier touch. See Xanadu 112.

7. In an informal performance (as yet unreleased) with Charlie Parker and Kenny Clarke in August 1951, Tristano showed that he *could* take a more interactive approach to comping.

8. *Descent into the Maelstrom*—1953, East Wind 8040 (unreleased until after Tristano's death).

9. *Line Up* and *East Thirty-second* are on Atlantic 1224; *All the Things You are* is on Elektra Musician 60264; *C Minor Complex* is on Atlantic 1357.

10. Many instrumentalists vocalize involuntarily while playing. Some are quiet and/or mellifluous in their vocalizing; Powell was neither.

11. *All God's Children Got Rhythm*—May 1949, Clef 11046; *Bud on Bach*—3 August 1957, Blue Note 1571.

12. Recorded by Carmen McRae in 1972, Atlantic 2-904.

13. See especially Auld's *In the Middle*—7 February 1945, Guild 116—and Parker's *Bird's Nest* and *Cool Blues*—19 February 1947, Dial (reissued on Spotlite 102).

14. On 14 March 1955 he recorded 19 pieces unaccompanied; they are scattered among several Mercury and EmArcy LPs.

15. This example is from Taylor 1982: 130 (edited).

16. 17 July 1952, MGM 11354.

17. 24 December 1947, Savoy 942.

18. He discusses the sensational aspects of his life in his autobiography Hawes & Asher 1972; for discographical information, see Hunter & Davis 1986.

19. The example is from the sixth chorus of *Soul Junction*—15 November 1957, Prestige 7181.

20. As a young man, Silver copied some musical ideas from recordings by boogie-woogie pianists Albert Ammons and Pete Johnson; see Kirchner 1976: 5.

21. 10 August 1959, Blue Note 4017.

22. *The Natives Are Restless Tonight*—16 April 1965, Emerald CD 1003.

23. He once told me that while comping he tries to "goose" his soloists; that is, he tries to push and prod them so they will play their best.

24. Some definitive blues solos by Silver include *Silver's Blue* (17 July 1956, Columbia 16006), *Señor Blues* (10 November 1956, Blue Note 1539), and *Everything's Gonna Be Alright* (25 August 1983, Silveto 103).

25. This excerpt is from *Sandy's Blues* (April 1968, MPS 15180), first chorus, measures 1–4.

26. 1968, MPS 15181. Of course, the fine pianos he plays help him produce an

exquisite tone. In recent years his performance contracts typically require concert halls—his high fee eliminates him from the jazz night-club circuit—to provide a Bösendorfer concert grand, one of the world's finest pianos.

27. This example is from the first chorus of *I Remember Clifford* (162, Verve 8681). Peterson's influence on other pianists is less than Bud Powell's, partly because of the great technical demands needed to play his characteristic runs and figures properly. But several younger pianists found some inspiration in Peterson's style: Monty Alexander, Hampton Hawes, Oliver Jones, Adam Makowicz, Tete Montoliu, Phineas Newborn, Jr., Tom Ranier, and André Previn.

28. 28 December 1959, Riverside 315.

29. 17 October 1956, RCA Victor 2534.

30. *Oleo*—9 September 1958, Columbia 32470; *Blue in Green* and *Flamenco Sketches*—March and April 1959, Columbia 8163.

31. Miles Davis initially received composer credit for *Blue in Green,* but Evans claims to have written the piece, using at the start two chords that Davis suggested (Keepnews 1988: 170).

32. 28 December 1959, Riverside 315.

33. 22 August 1962, Milestone 47066.

34. *Dark Eyes*—April 1964, Verve 8833; *Little Lulu*—Fall 1967, Verve 8727.

35. April 1977, Milestone 55003.

36. *Nai Nai, Pentecostal Feeling,* and *Free Form,* with Donald Byrd—11 December 1961, Blue Note 4118; *Hush* and *Shangri-La,* with Byrd—21 September 1961, Blue Note 4101.

37. March 1968, Blue Note 4279.

38. 17 March 1965, Blue Note 4195.

39. *Matrix*—26 February 1968, Solid State 18039; *Trio Music*—November 1981, ECM 2-1232.

40. 1981, Elektra Musician 60025.

Chapter Eight: Bassists and Drummers

1. Bullock & Douglas 1987, 20.

2. 23 December 1943, Signature 9001.

3. 2 April 1960, Fantasy 6015.

4. Apparently Harry Babasin established the precedent of using the cello in jazz; in December 1947 he recorded two pieces on that instrument during a Dodo Marmarosa session (Dial 208).

5. *I Can't Get Started*—9 July 1957, Jubilee 1054, and 16 January 1959, United Artists 4036; *Stormy Weather*—20 October 1960, Candid 8021; *Haitian Fight Song*—12 March 1957, Atlantic 1260; *Mingus Ah Um*—May 1959, Columbia 8171.

6. *One Bass Hit II*—9 July 1946, Musicraft 404; *Two Bass Hit*—22 August 1947, RCA Victor 20-2603.

7. August and September 1951, Dee Gee 1002. The sessions were reissued under the name of the Modern Jazz Quartet on Savoy 12046.

8. 1985, Concord 4315.

9. 29 April 1954, Prestige 7076.

10. See especially *Watergate Blues*—22 October 1975, Strata-East 19766, and *Yardbird Suite*—ca. 1978, Columbia 35573.

11. *The Seance*—Spring 1966, Contemporary 7621; *Nature Boy*—Galaxy 5127.

12. 14 July 1957, Blue Note 1569.

13. 23 Mary 1958, Prestige 7316.

14. 29 February 1960, Contemporary 7573.

15. He played several pieces on *Stars of Jazz,* a Los Angeles television show, with the Richie Kamuca Quintet on 7 April 1958.

16. 28 December 1959, Riverside 315.

17. Riverside 376 and 399 and Milestone 9125.

18. 21 December 1960, Atlantic 1364.

19. This example is from Herbie Hancock's recording of *Speak like a Child*—9 March 1968, Blue Note 4279.

20. 13 July 1977, JVC 6241.

21. For a fuller discussion of this evolution, see Theodore Dennis Brown's excellent and comprehensive dissertation (Brown, T. D. 1976), especially pp. 401ff., the source of most of my information on these pages.

22. Elvin Jones feels that "the cymbal is more related to the voice than anything else—they sing to you, it's like a choir . . ." Nolan 1977: 14.

23. Fall 1939, Hindsight 148 (radio broadcasts).

24. 29 April 1954, Prestige 7076.

25. 16 June 1954, Prestige 7059.

26. 23 May 1963, Blue Note 4146.

27. This example shows measures 25–29 of the first chorus of Charlie Parker's *Chasin' the Bird,* take 3—8 May 1947, Savoy 977.

28. This example is from the second chorus of the Clifford Brown/Max Roach recording, *Flossie Lou*—17 February 1956, Limelight 2-8201.

29. 5 August 1954, EmArcy 36008 (with Clifford Brown).

30. 30 December 1952, Clef 89129.

31. *Clifford's Axe*—April 1954, GNP 18; *Daahoud* (two takes)—6 August 1954, EmArcy 6075 and Mainstream 386; *Blues Walk* (two takes)—24 February 1955, EmArcy 36036 and 814 637-2.

32. A good visual documentaiton of his playing techniques is available in an hour-long concert performance video-recorded at Blues Alley in Washington, D.C. (1981, Jazz America).

33. 30 October 1958, Blue Note 4003.

34. 6 April 1953, Contemporary 3507.

35. 22 July 1952, Contemporary 3508.

36. An early example of asymmetric meter in jazz is Sahib Shihab's *Jamila* of 1957 (Savoy 12124); the theme, but not the solos, is in septuple meter.

37. A fine example of his brush work is in the Bud Powell recording of *John's Abbey*—May 1958, Blue Note 1598.

38. This example is from the middle of John Coltrane's solo in *Blue Train*—15 September 1957, Blue Note 1577 (Coltrane's session).

39. *Impressions*—3 November 1961, Impulse 42; *Your Lady*—18 November 1963, Impulse 50; *A Love Supreme*—9 December 1964, Impulse 77.

40. The lengthiest study of Elvin Jones is Kofsky's three-part series (1976–78).

41. 1976, Aebersold 1215.

Chapter Nine: Other Instrumentalists

1. Occasionally he doubles on soprano and alto saxophones, and once even traded instruments with Stan Getz for part of a recording session (12 October 1957, Verve 8249). Though he plays the smaller instruments with polish and professionalism, he has made his finest musical statements on the bari.

2. 1987, GRP 9544.

3. One of the great unanswerable questions in jazz is this: What would Artie Shaw's music have been had he not quit playing in 1954? That year he made a series of excellent small-group recordings with Hank Jones and other bebop players, and seemed to be moving toward the rhythmic and melodic norms of bebop.

4. Jazz at the Philharmonic recordings—2 July 1944, Clef Vols. 4, 5, 7; *Jay Bird, etc.*—26 June 1946, Savoy 2232; *Bone-o-logy*—24 December 1947, Savoy 2232.

5. These examples are from *Jerry Old Man,* with Henri Renaud—7 March 1954, Swing 33320, and *Blues for Trombones*—24 August 1954, Savoy 12010.

6. July 1988, Antilles 422-848 214-2 and 314-510 059-2. Baker (1979) and Bourgois III (1986) are useful studies of Johnson.

7. These examples are from recordings by the Modern Jazz Quartet: *Really True Blues* (27 April 1966, Atlantic 1468) and *Bluesology* (14 February 1956, Atlantic 1231).

8. *A New Sound—A New Star,* Blue Note 1512.

9. This example is from *James and Wes*—28 September 1966, Verve 8678.

10. This example is Montgomery's last solo chorus in *Somethin' Like Bags*—4 August 1961, Riverside 382.

11. For a discography and biography of Montgomery, see Ingram (1985).

12. Late 1961, Pacific Jazz 48.

13. December 1973, Pablo 2310 708.

14. He actually made his recording debut in 1954, while still a preteen, singing and playing on an obscure label.

15. *Breezin'*—1976, Warner Bros. 2919; *On Broadway*—1 February 1977, Warner Bros. 3139.

Chapter Ten: Ensembles

1. ensemble—"a system of items that constitute an organic unity: a congruous whole. . . ." (*Webster's Third New International Dictonary*).

2. May 1951, Blue Note 1503.

3. In 1987, Max Roach explained that after the initial attempt, Powell said, "Man, this ought to be Max Roach. Can't you think of nothing else to play?" (Brown, A. 1990: 54).

4. *Swinging on a Star*—September 1953, JATP Vol. 16; *Tangerine*—September 1962, Verve 8516.

5. For the first trio—Verve 8613 (January 1965); for the second trio—Warner Bros. 3504 (August 1977).

6. Steven Larson includes an extensive analysis and a full-score transcription of *'Round Midnight* from the first of these albums in his 1987 study of several performances of Thelonious Monk's famous composition.

7. *Conversations with Myself* (album)—January and February 1963, Verve 8526;

Further Conversations with Myself (album)—August and September 1967, Verve 8727; *My Romance*—20 January 1973, Fantasy 9457; *New Conversations* (album)—January and February 1978, Warner Bros. 3177; *Portrait in Jazz* (album)—28 December 1959, Riverside 315; *Turn Out the Stars*—January 1974, Fantasy 9501; Village Vanguard albums—25 June 1961, Riverside 376 and 399, and Milestone 9125.

8. March 1954, Columbia 566.

9. June 1954, Vogue 504 (= Pacific Jazz 1210).

10. In 1992, Kay suffered some medical problems that forced him to stop playing for several months; Mickey Roker took his place until spring 1993, when Kay rejoined the group.

11. *Django*—April 1960, Atlantic 1385; *Versailles*—22 January 1956, Atlantic 1231. For a full-score transcription of *Versailles* and a discussion of other pieces in the fugue series, see Owens (1976). The most extensive study of the MJQ is by Knauer (1990). Besides extensive analyses of the music, Knauer provides full-score transcriptions of several recordings, including some of the other fugues besides *Versailles.*

12. According to his son, Monk had collections of Chopin, Liszt, and other European piano composers in his library.

13. 31 October 1964, Columbia 38030.

14. *Blues Walk,* take 8—24 February 1955, EmArcy 36036; *Daahoud* and *Joy Spring*—6 August 1954, EmArcy 6075.

15. June 1952?, Jazztone 1231.

16. *Blues Walk,* take 5—24 February 1955, EmArcy 814 637-2. Brown had experimented with this exchange of progressively shorter phrases a few months earlier, in *Coronado,* another blues (EmArcy 36039), but the results were inferior to those in *Blues Walk.*

17. *At the Blackhawk, Vol. 1*—September 1959, Contemporary 7577. "West Coast jazz" was a term that not only implied a region, but a regional style, which was supposedly bland, subdued, and over-refined. Manne delighted in using the term sarcastically when introducing his band to audiences.

18. *All Bird's Children* (album)—June 1990, Concord 4441; *Gratitude* (album)—19 June 1986, Denon 1316; *Live from New York* (album)—7 October 1982, Palo Alto 8084; *The Phil Woods Quartet Volume One* (album)—26 May 1979, Clean Cuts 702.

19. *At the Lighthouse*—16 October 1960, Riverside 344.

20. *What Know*—14 September 1960, Blue Note 4054; *Pensativa* and *Free for All*—10 February 1964, Blue Note 4170; *Tell It Like It Is*—27 May 1961, Blue Note 4156.

21. 21 February 1954, Blue Note 1521, 1522, and 473-J2.

22. Recently he recorded *The Hillbilly Bebopper, The Walk Around-Look Up and Down Song,* and *Put Me in the Basement*—contrafacts of *Cherokee, In a Little Spanish Town,* and *Body and Soul*—February 1993, Columbia 53812.

23. *Accepting Responsibility*—September 1981, Silveto 101; *Activation*—1976, Blue Note 581-G; *Barbara*—1975, Blue Note 406-G; *Don't Dwell on Your Problems*—25 August 1983, Silveto 103; *Doodlin'*—13 November 1954, Blue Note 5058; *Everything's Gonna Be Alright*—25 August 1983, Silveto 103; *Expansion*—26 October 1979, Blue Note 1033; *Freeing My Mind*—1976, Blue Note 708G; *Gregory Is Here*—1969?, Blue Note 054-F; *Hangin' Loose*—31 March 1988, Silveto 105; *How Much Does Matter Really Matter*—1972?, Blue Note 4420; *Inner Feelings*—26 October 1979, Blue Note 1033; *Let the Music Heal Ya*—25 March 1985, Silveto 104; *Lonely Woman*—26 October 1964, Blue Note 4185; *Melancholy Mood*—10 August 1959, Blue Note 4017; *Moon*

Rays—3 January 1958, Blue Note 1589; *Music to Ease Your Disease*—31 March 1988, Silveto 105; *New York Lament*—1976, Blue Note 708G; *Nica's Dream*—9 July 1960, Blue Note 4042; *Nineteen Bars*—April 1963, Blue Note 4131; *Nutville*—22 October 1965, Blue Note 4220; *Optimism*—26 October 1979, Blue Note 1033; *Opus de Funk*—23 October 1953, Blue Note 5034; *The Outlaw*—3 January 1958, Blue Note 1589; *Peace*—8 April 1970, Blue Note 4352; *Perseverance and Endurance*—1976, Blue Note 581-G; *The Preacher*—6 February 1955, Blue Note 5062; *Señor Blues*—10 November 1956, Blue Note 1539; *Serenade to a Soul Sister*—29 March 1968, Blue Note 4277; *Song for My Father*—31 October 1964, Blue Note 4185; *Soul Searchin'*—1971?, Blue Note 4368; *Stop Time*—13 November 1954, Blue Note 5058; *There's No Need to Struggle*—25 August 1983, Silveto 103; *Time and Effort*—1976, Blue Note 581-G; *Who Has the Answer?*—1972?, Blue Note 4420; *Wipe Away the Evil*—18 June 1970, Blue Note 4352; *Won't You Open up Your Senses*—1971?, Blue Note 4368.

24. This casual approach worked well unless one or more of the players had only partial familiarity with the piece. Davis and Garland show an interesting disagreement on how the melody goes in *Surrey with the Fringe on Top* (1956), and then there are the mistakes in *When Lights Are Low* and *Well You Needn't* (pp. 116–17).

25. Spanish Phrygian is the Phrygian scale with both a raised and lowered third: D-E$^\flat$-F-F$^\#$-G-A-B$^\flat$-C-D. Kernfeld gives a convincing alternate interpretation of this piece's underpinning, saying that the structure is really a series of *chords,* not scales, in the key of C: CMA^7-A$^{\flat 7}$-B$^\flat$MA^7-D (embellished by E$^\flat$/$_\text{D}$)-GMI$^{(\text{ADD6})}$. See Kernfeld 1981: 137ff.

26. *All Blues* (see *Kind of Blue*); *Blue in Green* (see *Kind of Blue*); *Boplicity*—22 April 1949, Capitol 762; *Bye, Bye, Blackbird*—5 June 1956, Columbia 949; *Darn That Dream*—9 March 1950, Capitol 1221; *Flamenco Sketches* (see *Kind of Blue*); *Freddie Freeloader* (see *Kind of Blue*); *Kind of Blue* (album)—1959, Columbia 8163; *Miles Ahead* (album)—1957, Columbia 1041; *Moon Dreams*—9 March 1950, Capitol 762; *Porgy and Bess* (album)—1958, Columbia 8085; *'Round Midnight*—10 September 1956, Columbia 949; *Sketches of Spain* (album)—1959–60, Columbia 8271; *So What* (see *Kind of Blue*); *Surrey with the Fringe on Top*—11 May 1956, Prestige 7200; *Well You Needn't*—6 March 1954, Blue Note 5040; *When Lights Are Low*—19 May 1953, Prestige 7054.

27. Donald Johns showed me that the *a* section of Coltrane's *Impresssions* melody is a secondary theme in Morton Gould's *Pavanne* (1939). The *b* section perhaps derives from Maurice Ravel's *Pavane pour une infante defúnte,* measures 8ff.

28. 26 October 1963, Pablo 2620 101.

29. *Better Git It in Your Soul*—May 1959, Columbia 1370; *The Black Saint and Sinner Lady* (album)—20 January 1963, Impulse 35; *Duke Ellington's Sound of Love*—December 1974, Atlantic 1677; *Fables of Faubus*—5 May 1959, Columbia 1370; *Folk Forms No. 1*—20 October 1960, Candid 8005; *Goodbye Pork Pie Hat*—May 1959, Columbia 1370; *Haitian Fight Song*—12 March 1957, Atlantic 1260; *No Private Income Blues*—January 1959, United Artists 4036; *Pithycanthropus Erectus*—Fall 1970, America 30 AM 6109; *Pussy Cat Dues*—May 1959, Columbia 1370; *Wednesday Night Prayer Meeting*—4 February 1959, Atlantic 1305.

30. *Encore*—May 1963, Philips 600-092; *Big Band Goodies*—May 1963, Philips 600-171.

31. Recently Foster augmented the band's book with some pieces that are thoroughly in the bebop idiom.

32. *A Child Is Born*—25 May 1970, Blue Note 4346; *The Groove Merchant*—June 1969, Solid State 18058; *Kids Are Pretty People*—4 December 1968, Solid State 18048.

33. July 1956, EmArcy 36078. Merrill and Evans rerecorded 11 of the 12 pieces from the original album in August 1987, seven months before Evans died (EmArcy 834 205-2).

34. *Boplicity* and *Darn That Dream*—see Davis; *Great Jazz Standards* (album)—early 1959, World Pacific 1270; *The Individualism of Gil Evans* (album)—1963–64, Verve 8555 (Rudy Van Gelder recorded this album beautifully; he captured every nuance and unusual blend of instruments that Evans devised); *Miles Ahead* (album)—May 1957, Columbia 1041; *Moon Dreams*—see Davis; *New Bottles Old Wine* (album featuring Cannonball Adderley)—April-May 1958, World Pacific 1246; *Out of the Cool* (album)—November-December 1960, Impulse 4; *Porgy and Bess* (album)—July/August 1958, Columbia 1274; *Sketches of Spain* (album)—November 1959/March 1960, Columbia 8271.

35. April 1974, RCA (Japan) 1-0236, and 8 February 1976, RCA 2-2242.

36. *England's Carol*—June 1960, Atlantic 1369; *Piece for Clarinet and String Orchestra*—March 1960, Verve 8395; *Tangents in Jazz*—May 1955, Capitol 634.

37. 1986, GRP 9533.

38. Corea plays these linked-together synthesizers with a keyboard controller, which is a guitar-like device with three octaves of stubby keys for the right hand and pitch, volume, and timbre controls for the left.

39. *La Fiesta*—October 1978, ECM 1-1140; *Got a Match?*—1986?, GRP 9535.

Chapter Eleven: Younger Masters

1. Sancton 1990: 70.

2. Quoted in Feather 1991.

3. 1981, Columbia 37574.

4. *My Funny Valentine*—11 October 1980, Who's Who 21026; *The Jazz Messengers Big Band*—July 1980, Timeless 150.

5. 16 November 1955, Prestige 7254.

6. *Herbie Hancock Quartet*—1981, Columbia 38275.

7. In 1991 the Marsalis brothers recorded another interactive duet, *Cain and Abel* (Columbia 46990). The theme, with its unpredictable meter changes, frames 24-measure blues choruses in which the brothers engage in an almost continuous imitative contrapuntal conversation. It is a brilliant updating of the old New Orleans collective improvisation concept.

8. January 1982, Concord 196.

9. May 1984, Columbia 39530.

10. In a comment to a master class at El Camino College on 29 April 1988.

11. *Black Codes*—January 1985, Columbia 40009; *J Mood*—December 1985, Columbia 40308.

12. 1986, Columbia 40461.

13. *Standard Time Vol. 3: The Resolution of Romance*—1990?, Columbia 46143; *The Majesty of the Blues*—October 1988, Columbia 45091.

14. Recorded with the Mercer Ellington band—1986?, GRP 9548.

15. Ellis Marsalis has *four* musically talented sons, Branford, Wynton, Delfeayo, and Jason. In 1992, 15-year-old drummer Jason toured with Delfeayo's band and showed great promise; according to Wynton, he is the most talented member of the family!

16. Perhaps not coincidentally, Roberts won first prize in the first Thelonious Monk International Jazz Piano Competition, held in 1987.

17. *Live at Blues Alley* (with Wynton Marsalis)—December 1986, Columbia 40675; *Alone with Three Giants* (album)—1990, Novus/RCA 3109-2.

18. Petrucciani usually leads a trio, as did Evans.

19. He was born with osteogenesis imperfecta, a calcium deficiency that stunted his growth.

Glossary

***aaba* form** a formal design used in innumerable popular and jazz songs; the letters designate the two contrasting sections of the song.

action jazz Don Heckman's term for the music of Ornette Coleman, Cecil Taylor, Archie Shepp, and other post-bebop players.

Aeolian scale a seven-note scale also known as the natural minor scale; in the key of C the Aeolian scale is C-D-E$^\flat$-F-G-A$^\flat$-B$^\flat$-C.

alternate take a recording that the leader or record producer rejects during a recording session; sometimes alternate takes appear in posthumous anthologies years after the master takes first appear.

altissimo register the saxophone range that is above the official highest note of the instrument; notes in the altissimo register result from unusual fingerings and embouchures, and other special techniques.

arco Italian for "bow"; to play a string instrument with a bow.

arpeggio notes of a chord played consecutively, not simultaneously. The arpeggio may ascent or descend, and is usually rapid.

articulation the manner in which players begin and end individual notes, using lip slurs, various tonguings, touches, stick strokes, etc.

asymmetric meter five-beat, seven-beat, nine-beat (when subdivided 2 + 2 + 2 + 3), eleven-beat, etc. meters; these meters were extremely rare in jazz before the 1960s.

augmented interval an interval (such as a fourth, fifth, ninth, or eleventh) that is a half-step larger than that same interval as it occurs normally within the major scale; for example, C-G is a perfect fifth, C-G# is an augmented fifth.

augmented triad a three-note chord consisting of a major third and an augmented fifth, such as CEG#.

backbeat beats two and four in quadruple meter.

ballad a slow popular song, often in *aaba* form.

band an instrumental ensemble dominated by wind and percussion instruments. In jazz, bands range in size from quartets and quintets up to big bands of 16–18 players (sometimes more).

bars 1) the vertical lines used in music notation to delineate measures; 2) a synonym for measures.

big band an ensemble of two to five trumpets (four is most common), two to five trombones (usually three or four), four to six saxophonists (who may double on clarinets, flutes, and other woodwinds), and a rhythm section.

bitonality the effect produced by music played in two (or possibly more) keys simultaneously, or by two different triads sounding simultaneously.

block chords chord voicings on a keyboard instrument in which both hands play several notes simultaneously.

"blowing session" an unrehearsed performance, often for recording purposes, in which the soloists improvise as long as they wish.

blue note a note of varying intonation; to play or sing a blue note the musician uses portamento to rise and fall in pitch, often between the major and minor third and the major and minor seventh steps of a scale.

blues a family of twelve-measure (sometimes eight-, sixteen-, or twenty-four-measure) harmonic structures; the most common harmonic structures in jazz. In bebop, one of the favorite of these harmonic plans is |I |IV |I |I$^{\flat 7}$ |IV |IV |I |[V^7]ii |ii |V^7 |I[V^7]ii |ii V^7| (these basic triads and dominant sevenths are usually adorned with coloring tones, such as ninths and thirteenths).

blues scale a six-note scale consisting of notes 1, $\flat$3, 4, #4, 5, $\flat$7, and 8; in the key of C, the notes are C-E$\flat$-F-F#-G-B$\flat$-C.

bomb a fill played emphatically on the bass drum.

boogie-woogie solo piano blues with a constant eighth-note accompaniment pattern in the left hand and riff-dominanted melodies in the right.

bossa nova a Portuguese term for Brazilian songs composed as slow sambas, from the 1960s and after.

break a short unaccompanied solo, usually two or four measures long, and usually occurring in the middle or at the end of a chorus.

breath accent accenting a note with an extra surge of air only, without tonguing.

bridge the *b* section of a theme in *aaba* form; synonyms include "channel," "release," and "middle eight."

broken chord a chord played one note at a time.

cadence the end of a phrase, or the end of a piece.

cadential formula a short melodic or harmonic pattern used to end a phrase or a section of a piece.

cadenza an unaccompanied solo, often lengthy and in free meter, at the end of a piece.

cents a unit of measurement for musical intervals; an octave contains 1200 cents, a half-step 100, a whole step 200, etc.

"changes" short for "chord changes"; the series of chords that make up the harmonic structure of a theme.

chord three or more different pitches played simultaneously; often chords are formed by playing every note of a scale, such as the triads CEG or DFA in the major scale on C. The triad built up from the first note of the scale is labeled I (a major triad), the one built from the second note of the scale is labeled ii (a minor triad), then follow the iii triad, the IV triad, the V triad, and others.

chord progression a series of different chords; one of the most common in jazz is ii-V-I.

chord substitution replacing a chord in a progression with a different but related chord.

chorus a single playing of a theme or of one improvisation on the chords of a theme. Each chorus of the blues *Now's the Time* is 12 measures long; each chorus of *Oleo* is 32 measures long.

chromatic descent moving down by half-steps through part of the chromatic scale.

chromatic motion moving up or down by half-steps through part of the chromatic scale.

chromatic scale a scale that includes all twelve pitches in the Western tuning system.

close position an arrangement of the notes of a chord within one octave or less, such as root-third-fifth or root-third-fifth-seventh; the opposite of open position.

cluster simultaneous playing of several adjacent scale tones.

coda the ending added after the final chorus of a performance.

codetta a short extension of a chorus, such as the extra two measures at the end of the original chorus of *I Got Rhythm.*

collective improvisation all the melody instruments of a group improvising simultaneously.

comping short for "accompanying"; the playing of chords in intermittent, irregular rhythms as an accompaniment for a soloist.

consonance a stable, restful sound, such as is produced by the intervals in major and minor triads.

contrafact short for melodic contrafact; a new melody composed upon the chord changes of a pre-existing piece.

cool jazz a label often applied to subdued bebop performances.

countermelody a subsidiary melody played concurrently with a theme or with an improvised solo melody.

crash cymbal a small-diameter cymbal that produces a more sudden and higher splash of sound than that produced by a ride cymbal.

crescendo a gradual increase in loudness.

decay the gradual fading of sound produced by a struck or plucked instrument—drum, cymbal, piano, plucked string bass, etc.

decrescendo a gradual decrease in loudness.

develop to base one or more phrases on manipulations of a single motive.

diatonic referring to melodies and harmonies based on a seven-note scale; the opposite of **chromatic.**

diminished fifth a fifth that is a half-step smaller than a perfect fifth; C-G is a perfect fifth, and C-G♭ is a diminished fifth.

diminished interval an interval that is a half-step smaller than a minor or perfect interval.

diminished scale an eight-note scale consisting of half-steps alternating with whole steps.

diminished seventh chord a four-note chord whose notes subdivide the octave into four equal intervals.

diminished triad a three-note chord consisting of a minor third and a diminished fifth; the C diminished triad is CE♭G♭.

dissonance an unstable, restless sound, such as is produced by playing adjacent scale tones simultaneously.

dominant the fifth step of a seven-note scale; also, a chord built on that step.

dominant seventh chord a four-note chord built on the fifth step of the scale; one of the most basic chords in Western harmony.

Dorian scale a seven-note scale similar to the natural minor scale; in the key of C the Dorian scale is C-D-E♭-F-G-A-B♭-C.

double stop two notes played simultaneously on a violin, viola, cello, or string bass.

downbeat the first beat of a measure.

dynamics the loudness levels of music.

eighth note the symbol (♪ and ♫) for notes that each last half a beat in most jazz.

eleventh a large interval spanning an octave and a fourth.

embouchure the position of the lips of a wind instrumentalist.

enriched dominant a dominant seventh chord with additional notes, such as an augmented fifth (replacing the perfect fifth), a minor ("flat") ninth, an augmented eleventh, or a thirteenth.

even eights eighth notes played evenly in time, dividing the beat into two equal halves.

exposition the initial portion of a fugue in which each instrument plays the subject at least once.

fall-off a rapid, descending portamento at the end of a note, produced on a wind instrument by a lip gliss.

fifth 1) an interval five scale steps in size (for example, C-G or A-E in the major scale on C); 2) the note five scale steps higher than the root of a chord.

fills drum strokes, notes, chords, or short melodic figures that accompanists play while comping.

flatted fifth the common jazz term for **diminished fifth.**

forte, fortissimo instructions to play "loud," "very loud."

fours same as **trading fours.**

fourth chords chords built with fourths (such as CFB♭) instead of with thirds (such as CEG).

free jazz the term most often used in jazz writing for the music of Ornette Coleman, Cecil Taylor, Archie Shepp, and other post-bebop players.

free meter a meter in which the tempo and/or number of beats per measure is constantly changing.

front line the wind instruments in a group.

fugue a composition based on a short melodic fragment—the subject—in which the instrumental parts are rhythmically independent and equally active; a rare genre in jazz.

funky jazz bebop flavored with borrowings from the folk-blues and gospel-music traditions.

fusion jazz a jazz style that blends elements of jazz and rock.

grace note a quick note used to embellish the beginning of a longer note.

half-diminished seventh chord a four-note chord consisting of a diminished triad and the interval of a minor seventh; C half-diminished seventh is CE♭G♭B♭.

half-note a symbol (𝅗𝅥) for a note lasting two beats in most jazz.

half-step the smallest interval on a keyboard instrument.

half-valve pushing the valves of a trumpet or cornet down only partway to produce an unusual pitch and tone quality.

harmonic minor scale see **minor scale.**

Harmon mute a metal mute, with an adjustable stem in the center, for trumpet (and less commonly for trombone).

harmony the chords used in music.

head arrangement an unwritten arrangement, arrived at collectively by the players.

heterophony simultaneous playing of two or more slightly different versions of the same melody.

high hat a pair of cymbals mounted on a stand and operated with a foot pedal.

Hz Hertz, the unit of frequency of a vibrating acoustical body, equal to one complete vibration cycle. The A above middle C has a frequency of 440 Hz.

interval the distance between two notes. We measure an interval's size by counting the letter names from the lower to the upper note: A-B is a second, A-C a third, A-E a fifth, A-A an eighth (more commonly an octave), etc.

intonation the tuning of one note relative to another; some wind instrumentalists readily adjust their intonation to match the pitches of a piano, others do not.

inverted mordent a rapid three-note figure that ascends and then descends by step (♪♪♪ or ♪♬).

jam session an impromptu, and often private, performance by a loosely organized group of players.

key 1) the levers on keyboard and woodwind instruments that players press; 2) the pitch center, or tonic, around which the melodies and harmonies of a piece revolve.

lay out to stop playing while others continue to play.

lead sheet a concisely written composition, consisting of the melody with chord symbols superimposed.

legato to connect notes together smoothly, without a break in the sound.

lick a melodic phrase that a player uses regularly in improvising.

lip gliss or **lip slur** a quick rise or fall of pitch on a wind instrument, produced by tightening or relaxing the lip and/or by changing wind pressure.

locked hands a piano technique in which the hands work together rhythmically. The right hand plays a harmonized melody and the left hand doubles the melody an octave lower.

lowered fifth, ninth synonyms for diminished fifth and minor ninth; intervals that are a half-step smaller than a perfect fifth and major ninth—C-$G^{\flat}$ and C-$D^{\flat}$, for example.

lower neighbor tone a note a half- or whole step lower than a chord tone.

major intervals seconds, thirds, sixths, and sevenths in which the upper note is in the major scale of the lower note; C-E, C-A, and C-B are respectively a major third, major sixth, and major seventh.

major scale an important seven-note scale in Western music; in the key of C the major scale is C-D-E-F-G-A-B-C.

major triad a triad formed by the first, third, and fifth steps of a major scale (such as CEG in the major scale on C).

manual on an organ, a keyboard for the hands.

master take the version of a recording that the band leader and/or record producer selects to be issued.

measure a grouping of beats equal to the metric pattern; in triple meter a measure contains three beats, in quadruple meter a measure contains four beats, etc.

melodic contrafact a new melody composed upon the chord changes of a pre-existing piece.

melodic minor scale see **minor scale.**

meter a pattern of regularly recurring groups of strong and weak beats, such as strong-weak-weak in triple meter, or strong-weak-strong-weak in duple or quadruple meter.

mezzo forte an instruction to play "medium soft."

minor intervals intervals that are a half-step smaller than major intervals; C-D$^{\flat}$, C-E$^{\flat}$, C-A$^{\flat}$, and C-B$^{\flat}$ are respectively a minor second, minor third, minor sixth, and minor seventh.

minor scale any one of a set of three important scales in Western music; in the key of C the natural minor scale is C-D-E$^{\flat}$-F-G-A$^{\flat}$-B$^{\flat}$-C, the harmonic minor scale is C-D-E$^{\flat}$-F-G-A$^{\flat}$-B-C, the melodic minor scale is C-D-E$^{\flat}$-F-G-A-B-C-B$^{\flat}$-A$^{\flat}$-G-F-E$^{\flat}$-D-C.

minor triad a triad formed by the first, third, and fifth steps of a minor scale (such as CE$^{\flat}$G in the minor scale on C).

Mixolydian scale a scale similar to the major scale, but with the seventh lowered; in C the Mixolydian scale is C-D-E-F-G-A-B$^{\flat}$-C.

modal jazz a term used for jazz that incorporates improvisation on modes (or scales) instead of on chord progressions.

mode a term often used as a synonym for scale.

monophony the playing of a single melody without accompaniment.

motive a short melodic or rhythmic fragment, used repeatedly, but not regularly and constantly, in a phrase, in several phrases, or even throughout a solo or composition.

natural overtone series the pitches whose frequencies are whole-number multiples of the frequency of a sounded note; for the low A on a guitar (110 Hz), the natural overtone series begins with the pitches A, E, A, C#, and E (220 Hz, 330 Hz, 440 Hz, 550 Hz, and 660 Hz), all of which are present in the sound of the plucked A string.

ninth a large intreval spanning an octave and a second; on C a major ninth is C-D, an augmented (or "raised" or "sharp") ninth is C-D#, and a minor (or "lowered" or "flat") ninth is C-D♭.

Noh a Japanese musical-theatrical genre that dates from the 14th century; music for Noh dramas is provided by a wooden flute, three drums, and voices.

octave an interval eight scale steps in size; both notes have the same letter name.

open position an arrangement of chord tones that spans more than one octave.

orchestra a large instrumental ensemble dominated by bowed string instruments. Orchestras are rare in jazz (though in the jazz world the term is a common synonym for band), except on special occasions when jazz musicians perform with symphony orchestras.

ostinato a short fragment of music repeated regularly and constantly throughout a piece or large section of a piece.

"outside" playing to improvise using primarily notes that are not in the chord or scale of the moment; common in later bebop and in post-bebop styles.

parallel thirds or sixths the playing of two rhythmically identical melodic lines, where the subsidiary melody is consistently a third or a sixth below the main melody.

parallel octaves playing a melody in octaves rather than single pitches. When keyboardists play the same melody in each hand and the hands are two octaves apart, they are playing parallel fifteenths (or parallel double octaves).

passing chord a decorative chord whose root connects the roots of two structural chords by whole step or by half-step; in the progression C_{MI}^{7}-$\text{C}\#_{\text{MI}}^{7}$-$\text{D}_{\text{MI}}^{7}$, the middle chord is a passing chord.

pedal tone a sustained or repeated bass note played throughout a large section of a musical piece.

pentatonic scale any of several five-note scales; the major-pentatonic consists of notes 1, 2, 3, 5, and 6 of a major scale.

perfect fourth, perfect fifth the intervals of the fourth and fifth in which each note is in the major scale of the other; on C, C-F and C-G are respectively a perfect fourth and a perfect fifth.

phrasing the manner in which the notes of a phrase are played, as in legato phrasing, or staccato phrasing.

Phrygian scale a scale similar to the natural minor scale, but with a lowered second step; in C the Phrygian scale is C-D♭-E♭-F-G-A♭-B♭-C.

piano, pianissimo instructions to play "soft," "very soft."

pickup 1) a synonym for upbeat—the note or notes that precede the downbeat of a piece or of a phrase; 2) a small microphone attached to an instrument.

pizzicato plucked; the usual way of playing the string bass in jazz.

polyphony literally "many sounds"; music in which more than one pitch sounds simultaneously, either in chords or in two or more simultaneous melodies.

popular standards popular songs—especially those from the 1920s through the 1950s that George Gershwin, Jerome Kern, Richard Rodgers, and others composed—that have remained in the repertoire; many are 32 measures long and in *aaba* or *abac* form.

portamento moving from one note to another by sliding through all possible pitches in between.

press roll in drumming, a series of double-stroke rolls (LLRR-LLRR . . .).

quartatonic scale a four-note scale.

quarter note the symbol (♩) for a note lasting one beat in most jazz.

raised fifth, raised ninth see **ninth.**

range the total span of pitches playable on an instrument.

register a portion of an instrument's range, such as low register or high register.

rhythm section the instruments that are played nearly all the time during a performance: most often piano, bass, and drums, but guitar, organ, other keyboard instruments, and uncommon (for jazz) percussion instruments also belong in this section.

ride cymbal a large suspended cymbal used extensively by drummers to accompany soloists.

riff a short phrase (usually two or four measures long) that repeats several times.

rim shot a stroke of the drum stick on the center of the drum head and on the metal counter-hoop (rim) simultaneously.

ritornello a short section that recurs periodically throughout a composition.

roll rapid repetitions of a single note on a drum or xylophone.

root the lowest note of a chord arranged in thirds (in the chord CEG, C is the root).

rhythm guitar a guitar used to strum chords softly on each beat.

"running the changes" to base one's solo almost entirely on the notes of the chords in the chorus; the term is often derogatory, for the resulting melodies often sound mechanical and unvocal.

sarod an important plucked string instrument of North India.

saxello a rare member of the saxophone family; similar in tone and range to a soprano saxophone.

scalar pertaining to a scale.

scalar improvisation improvisation based on a single scale for extended sections; a distinctively different premise from improvisation on chord progressions.

scale 1) the set of pitches, starting with the tonic and arranged in ascending or descending order, that is used in a piece of music; 2) wages established by the American Federation of Musicians for different types of performances.

scat singing using vocables to sing improvised jazz melodies.

second a small interval, usually either a half-step (minor second) or a whole step (major second).

secondary dominant a dominant seventh chord built on a scale step other than the dominant; in the key of C, G^7 is the primary dominant, and C^7, D^7, E^7, F^7, and A^7 are secondary dominants. Just as G^7 normally progresses to a chord on C, a fifth lower, so each secondary dominant usually progresses to a chord whose root is a fifth lower.

sequence repetition of a melodic fragment or chord progression at a higher or lower pitch.

seventh an interval spanning seven scale steps; on C, a major seventh is C–B, a minor seventh is C–B♭, a diminished seventh is C–B♭♭.

"sheets of sound" a term Ira Gitler coined to describe the barrages of notes that John Coltrane often played while improvising.

shuffle beat a lively rhythm pattern having a pair of swing eighths on each beat.

sidelining pretending to play an instrument, usually during the making of a television show or film.

sideman any player in a group other than the leader.

sine wave the pattern formed by a pure tone; that is, a tone with no overtones, only a fundamental pitch.

sixteenth note a symbol (𝅘𝅥𝅯 or ♬) for notes that each last one-fourth of a beat in most jazz.

sixth an interval spanning six note names; C–A is a major sixth, C–A♭ is a minor sixth.

slap-tonguing a percussive effect produced by saxophonists.

snare drum a small, flat drum with a series of metal spirals (snares); the snares rattle against the lower drumhead when the upper head is struck, but a lever gives the player the option of releasing the tension on the snares to eliminate the rattling.

"Spanish Phrygian" a term sometimes used to label a Phrygian scale that has both a major and a minor third; in C the "Spanish Phrygian" scale is C-D♭-E♭-E-F-G-A♭-B♭-C.

staccatissimo a shorter articulation of notes than *staccato*.

staccato to play a melody with silences between successive notes.

stretto a section of a fugue in which the various instruments play overlapping statements of the subject.

stride piano style to play low-register and mid-register notes alternately in the left hand, thus producing a "boom-chick-boom-chick" pattern.

structural tone an important chord tone (usually the root, third, or fifth) in an important chord in the harmony of a song; in *Bye, Bye, Blackbird,* the first melody note—the third of the tonic chord—is a structural tone.

subject a short melodic fragment used as the basis of a fugue.

swing 1) a verb referring to the rhythmic aspects of jazz; for an ensemble to swing the rhythm section must synchronize its playing of the beats and the subdivisions of the beat, and the soloist must both synchronize with the rhythm section and create a sense of forward momentum using a skillful manipulation of phrasings and articulations; 2) the style of jazz (also called big-band jazz) that preceded bebop in the 1930s.

swing eighths two unevenly spaced notes per beat, with the first note sounding at the beginning of the beat and the second one sounding on the last third of the beat (♩ 𝄾 ♩ (triplet, 3) or ♩ ♪ (triplet, 3)).

syncopation to play accented notes between beats or on metrically weak beats.

tabla a pair of hand drums in North India.

take a recorded performance, whether complete or incomplete, of a piece; often a group will record several takes before producing one acceptable as the master take.

tambura a four-string drone instrument in North India.

tenth an interval spanning an octave plus a third; it is the largest interval that most keyboardists can reach comfortably.

terminal vibrato a brief oscillation used at the end of a sustained note on a wind instrument.

tertian harmonies harmonies, or chords, built with thirds.

texture the layers of sound in music.

third an interval spanning three consecutive notes in a scale (such as C-E or D-F); also, the note three scale steps higher than the root of a chord (in the chord CEG, E is the third).

thirteenth a large interval, spanning an octave and a sixth; it is a common coloring tone added to a dominant seventh.

thirty-second note a symbol (𝅘𝅥𝅰 or ♬) for quick notes that each last only an eighth of a beat in most jazz.

through-composed a term denoting a theme or composition that contains no large-scale repetition.

timbre tone color; the way a musical instrument sounds.

toccata a fast composition designed to showcase the speed and virtuosity of the player.

tonality a pitch center, or musical "center of gravity," usually expressed in terms of "key," such as "in the key of B♭."

tonic the pitch center, or keynote, of a piece; for a piece in the key of C, the note C is the tonic.

tonic chord one of several chords built on the tonic note; in the key of C major, the tonic chords include CEG and CEGB.

trading fours, trading eights players taking turns playing four- or eight-measure phrases; often occurs between the wind player(s) and the drummer in a group.

tremolo a rapid oscillation between two notes a third or more apart, especially on a keyboard instrument, xylophone, or vibraphone.

trill a rapid oscillation between two notes a half-step or whole step apart.

triplet a group of three notes equally spaced in time and filling the time normally occupied by two notes: triplet eighth notes fill one beat, the time filled by two standard eighth notes.

tritone an interval spanning three whole steps, and dividing the octave into two equal halves.

tritone substitution the replacement of a chord with a similar sounding chord whose root is a tritone away from the original chord; the progression G^7-C becomes $D^{\flat 7}$-C after tritone substitution.

turnaround a short series of chords, usually lasting two or four measures, inserted at the end of a chorus; turnarounds pull the music forward to the beginning of the next chorus.

tutti all; the full band playing a passage.

two-beat emphasizing the strong beats in $\frac{4}{4}$ meter, by playing mostly half notes, instead of emphasizing all four beats.

two-part, three-part two simultaneous melodic lines, three simultaneous melodic lines.

unison two or more instruments playing the same melody simultaneously and in the same octave.

vamp a chordal ostinato.

vibrato a wavering pitch, usually spanning about a semitone and oscillating about six times per second; depending on the instrument and the player, vibrato may result from small and regular movements of the finger, hand, lips, or larynx.

voice leading the rhythmic and intervallic relationships between two or more simultaneous melodic lines.

voicing the distribution of chord tones.

walking bass line a bass line that moves by step, one note per beat, most of the time.

whole step an interval formed by two half-steps.

wholetone scale a scale made up entirely of whole steps; there are only two wholetone scales in the Western tuning system: C-D-E-F#-G#-A#-C and D♭-E♭-F-G-A-B-D♭.

Bibliography

Allen 1981

Daniel Allen. *Bibliography of Discographies, Volume 2: Jazz.* New York: Bowker.

Anonymous 1955

Anonymous. "Unknown Gets Big Jazz Date," *Down Beat* 22 (24 August), 16.

Astrup 1984

Arne Astrup. *The Revised Stan Getz Discography.* Karlslunde, Denmark: Meistrup.

Baker 1978a

David Baker. *John Coltrane: Tenor Saxophone.* New York: Hansen.

Baker 1978b

David Baker. *Miles Davis: Trumpet.* New York: Hansen.

Baker 1979

David Baker. *J. J. Johnson, Trombone.* New York: Hansen.

Baker 1985–86

David Baker. *How to Play Bebop.* Van Nuys, Calif.: Alfred. 3 vols.

Balliett 1959

Whitney Balliett. *The Sound of Surprise.* New York: Dutton.

Berendt 1959

Joachim Berendt. *Das neue Jazzbuch.* Frankfurt: Fischer.

Berger et al. 1982

Morroe Berger, Edward Berger, and James Patrick. *Benny Carter: A Life in American Music.* Metuchen, N.J.: Scarecrow Press. 2 vols.

Bittan 1959

David B. Bittan. "Don't Call Me Bird!" *Down Beat* 26 (14 May), 19–20.

Blake 1982

Ran Blake. "The Monk Piano Style," *Keyboard* 8 (July), 26ff.

Blancq 1977

Charles Clement Blancq. "Melodic Improvisation in American Jazz: The Style of Theodore 'Sonny' Rollins, 1951–62." Ph.D. dissertation, Tulane University.

Blancq 1983

Charles Clement Blancq. *Sonny Rollins: The Journey of a Jazzman.* Boston: Twayne.

Bourgois 1986

Louis George Bourgois III. "Jazz Trombonist J. J. Johnson: A Comprehensive Discography and Study of the Early Evolution of His Style." DMA dissertation, Ohio State University.

Brandt and Roemer 1976

Carl Brandt and Clinton Roemer. *Standardized Chord Symbol Notation (A Uniform System for the Music Profession).* Sherman Oaks, Calif.: Roemer.

Britt 1989

Stan Britt. *Long Tall Dexter.* London: Quartet; Reprinted in 1989 as *Dexter Gordon, a Musical Biography.* New York: Da Capo.

Brofsky 1983

Howard Brofsky. "Miles Davis and *My Funny Valentine:* The Evolution of a Solo," *Black Research Journal* (1983), 23ff.

Brown, A. 1990

Anthony Brown. "Modern Jazz Drumming Artistry," *Black Perspective in Music* 18 (1990), 39ff.

Brown, J. R. 1987

John Robert Brown, transcriber. *Jazz Trumpeter 2: Featuring Miles Davis.* London: Warner Bros.

Brown, T. D. 1976

Theodore Dennis Brown. "A History and Analysis of Jazz Drumming to 1942." Ph.D. dissertation, University of Michigan.

Bruyninckx 1984a

Walter Bruyninckx. *Modern Jazz* [Discography]: *Be-Bop, Hard Bop, West Coast.* Mechelen, Belgium: Bruyninckx. 6 vols.

Bruyninckx 1984b

Walter Bruyninckx. *Progressive Jazz* [Discography]: *Free, Third Stream, Fusion.* Mechelen, Belgium: Bruyninckx. 5 vols.

Bruyninckx 1985

Walter Bruyninckx. *Modern Jazz* [Discography]: *Modern Big Band.* Mechelen, Belgium: Bruyninckx. 2 vols.

Bullock & Douglas 1987

Paul Bullock and Bob Douglas. "Rapping with Red Callender," *Jazz Journal* (Los Angeles) 7 (Oct.–Dec.), 14ff.

Carles et al. 1988

Philippe Carles, André Clergeat, Jean-Louis Comolli. *Dictionnaire du Jazz.* Paris: Laffont.

Carner 1990

Gary Carner, compiler. *Jazz Performers: An Annotated Bibliography of Biographical Materials.* New York: Greenwood.

Carr 1982

Ian Carr. *Miles Davis, a Biography.* New York: Morrow.

Chambers 1983
Jack Chambers. *Milestones I: The Music and Times of Miles Davis to 1960.* Toronto: University of Toronto.
Chambers 1985
Jack Chambers. *Milestones II: The Music and Times of Miles Davis Since 1960.* Toronto: University of Toronto.
Chilton 1990
John Chilton. *The Song of the Hawk: The Life and Recordings of Coleman Hawkins.* Ann Arbor: Univ. of Michigan.
Coltrane 1960
John Coltrane, with Don DeMichael. "Coltrane on Coltrane," *Down Beat,* 27(29 Sept.), 26–27; reprinted in *Down Beat,* 12 July 1979.
Coolman 1985
Todd Coolman. *The Bass Tradition.* New Albany, Indiana: Aebersold.
Cuscuna 1984
Michael Cuscuna. *The Blue Note Story.* New York: Manhattan Records (sleeve notes for many Blue Note reissues).
Darroch 1984
Lynn Darroch. "Phil Woods: Bop Is Forever," *Jazz Times,* (Jan.), 10–12.
Detro 1988
John Detro. "Sonny Criss Fund Set—Hard Truth Revealed by Sonny's Mother," *JazzTimes* (Aug.), 6.
DeVeaux 1988
Scott DeVeaux. "Bebop and the Recording Industry: The 1942 AFM Recording Ban Reconsidered," *Journal of the American Musicological Society* 41 (Spring), 126ff.
Diaz 1991
Vince Diaz. "A Fleeting Glimpse of Stan Getz in Mexico," *L.A. Jazz Scene* 47 (July), 7.
Ellison 1964
Ralph Ellison. *Shadow and Act.* New York: Random House.
Feather 1957
Leonard Feather. *The Book of Jazz.* New York: Meridian.
Feather 1960
Leonard Feather. *The New Edition of the Encyclopedia of Jazz.* New York: Bonanza.
Feather 1991
Leonard Feather. "The Future? Well, You Needn't Worry," *Los Angeles Times* (Dec. 1), Calendar section, 67.
Fox 1959
Charles Fox. "Gil Evans: Experiment with Texture," in Raymond Horricks, ed., *These Jazzmen of Our Time.* London: Gollancz, 93ff.
Giddins 1981
Gary Giddins. *Riding on a Blue Note.* New York: Oxford.

Giddins 1987

Gary Giddins. *Celebrating Bird: The Triumph of Charlie Parker*. New York: Morrow.

Gillespie/Fraser 1979

Dizzy Gillespie and Joe Fraser. *to BE, or not . . . to BOP*. New York: Doubleday.

Gitler 1966

Ira Gitler. *Jazz Masters of the '40s*. New York: Macmillan.

Gitler 1985

Ira Gitler. *Swing to Bop*. New York: Oxford.

Gridley 1991

Mark C. Gridley. *Jazz Styles*. 4th ed. Englewood Cliffs, N.J.: Prentice-Hall.

Hawes & Asher 1972

Hampton Hawes and Don Asher. *Raise Up off Me*. New York: Coward, McCann & Geoghegan.

Hellhund 1985

Herbert Hellhund. *Cool Jazz*. Mainz: Schott.

Hentoff 1955

Nat Hentoff. "Miles," *Down Beat* 22 (2 Nov.), 13–14.

Hinton 1988

Milt Hinton. *Bass Line*. Philadelphia: Temple Univ. Press.

Hodeir 1956

André Hodeir. *Jazz, Its Evolution and Essence*. Translated by David Noakes. New York: Grove.

Hodeir 1962

André Hodeir. *Toward Jazz*. Translated by Noel Burch. New York: Grove.

Hunter & Davis 1986

Roger Hunter & Mike Davis. *Hampton Hawes: A Discography*. Manchester, Engl.: Manyana.

Ingram 1985

Adrian Ingram. *Wes Montgomery*. Gateshead, Engl.: Ashley Mark.

Jepsen 1963–70

Jorgen Grunnet Jepsen. *Jazz Records 1942–196_*. Holte, Denmark: Knudsen. 8 vols. in 11 parts.

Jones 1987

Max Jones. *Talking Jazz*. London: Macmillan.

Jost 1974

Ekkehard Jost. *Free Jazz*. Graz: Universal.

Julien 1984

Ivan Julien, transcriber. *Miles Davis "en Concert."* Paris: Alain Pierson.

Keepnews 1988

Orrin Keepnews. *The View from Within: Jazz Writings, 1948–1987*. New York: Oxford.

Kernfeld 1981
Barry Kernfeld. "Adderley, Coltrane, and Davis at the Twilight of Bebop: The Search for Melodic Coherence (1958–59)." Ph.D. dissertation, Cornell University.
Kernfeld 1988
Barry Kernfeld, ed. *The New Grove Dictionary of Jazz*. London: Macmillan. 2 vols.
Kirchner 1976
Bill Kirchner. "The Enduring Silver Standard," *Radio Free Jazz* 17 (Oct.), 4ff.
Kirchner 1978
Bill Kirchner. "Lee Konitz Today," *Radio Free Jazz* 19 (Sept.), 10–11.
Knauer 1990
Wolfram Knauer. *Zwischen Bebop und Free Jazz: Komposition und Improvisation des Modern Jazz Quartet*. Mainz: Schott. 2 vols.
Koch 1988
Lawrence O. Koch. *Yardbird Suite: A Compendium of the Music and Life of Charlie Parker*. Bowling Green, Ohio: Bowling Green Univ. Popular Press.
Kofsky 1970
Frank Kofsky. *Black Nationalism and the Revolution in Music*. New York: Pathfinder.
Kofsky 1976–78
Frank Kofsky. "Elvin Jones," *Journal of Jazz Studies* 4 (Fall 1976), 3ff.; 4 (Spring/Summer 1977), 11ff.; 5 (Fall/Winter 1978), 81ff.
Koster & Bakker 1974–76
Piet Koster & Dick M. Bakker. *Charlie Parker* [discography]. Alphen aan den Rijn, Holland: Micrography. 4 vols.
Koster & Bakker 1985, 1988
Piet Koster & Dick M. Bakker. *Dizzy Gillespie* [discography]. Amsterdam: Micrography. 2 vols.
Kynaston 1992
Trent Kynaston. "Ben Webster's Solo on 'Cotton Tail'—a Tenor Saxophone Transcription," *Down Beat* 59 (March), 59–60.
Larson 1987
Steven Leroy Larson. "Schenkerian Analysis of Modern Jazz." Ph.D. dissertation, University of Michigan. 3 vols.
Lees 1983
Gene Lees. "Paul," *Jazzletter* 3 (Oct.) 1.
Lees 1988
Gene Lees. *Oscar Peterson*. Toronto: Lester & Orpen Dennys.
Lees 1989
Gene Lees. "Waiting for Dizzy, Parts I & II," *Jazzletter* 8 (Aug. & Sept.).

Lees 1990

Gene Lees. "The Good Gray Fox, Part One," *Jazzletter* 9 (Nov.), 2ff.

Leiter 1953

Robert D. Leiter. *The Musicians and Petrillo.* New York: Bookman.

Levin & Wilson 1949

Michael Levin and John S. Wilson. "No Bop Roots in Jazz: Parker," *Down Beat,* 16 (9 September), 1ff.

Lyons 1976

Len Lyons. "New Intuitions: Bill Evans," *Down Beat* 43 (11 March), 12 ff.; reprinted in Len Lyons. *The Great Jazz Pianists, Speaking of Their Lives and Music.* New York: Morrow, 1983.

Mansfield and Green 1984

Horace Mansfield, Jr., and John Green. "Jackie McLean," *Be-Bop and Beyond* 2 (Sept.–Oct.), 14–20.

Miller 1989

Mark Miller. *Cool Blues: Charlie Parker in Canada, 1953.* London, Ontario: Nightwood.

Morgan & Mack 1987

Frank Morgan and Gerald Mack. "Be-Bop Lives: Frank Morgan Interviewed," *Be-Bop and Beyond* 5 (Summer), 13–16.

Nakagawa 1975

Kenji Nakagawa, transcriber. *Miles Davis: Jazz Improvisation.* Tokyo: Nichion. 2 vols.

Nolan 1977

Herb Nolan. "Rhythmic Pulsemaster: Elvin Jones," *Down Beat* 44 (15 Dec.), 13ff.

Owens 1974a

Thomas Owens. "Applying the Melograph to 'Parker's Mood,'" *Selected Reports in Ethnomusicology* 2, pp. 167–75.

Owens 1974b

Thomas Owens. "Charlie Parker: Techniques of Improvisation." Ph.D. dissertation, University of California at Los Angeles. 2 vols.

Owens 1976

Thomas Owens. "The Fugal Pieces of the Modern Jazz Quartet," *Journal of Jazz Studies* 4 (Fall), 25ff.

Patrick 1975

James Patrick. "Charlie Parker and Harmonic Sources of Bebop Composition: Thoughts on the Repertory of New Jazz in the 1940s," *Journal of Jazz Studies* 2 (June), 3ff.

Pepper and Pepper 1979

Art and Laurie Pepper. *Straight Life: The Story of Art Pepper.* New York: Schirmer.

Perlman & Greenblatt 1981

Alan M. Perlman and Daniel Greenblatt. "Miles Davis Meets Noam Chomsky: Some Observations on Jazz Improvisation and Language

Structure," in Wendy Steiner, ed., *The Sign in Music and Literature*. Austin: Univ. of Texas, 169ff.

Porter 1983
Lewis Porter. "John Coltrane's Music of 1960 Through 1967: Jazz Improvisation as Composition." Ph.D. dissertation, Brandeis University.

Porter 1985
Lewis Porter. *Lester Young*. Boston: Twayne.

Porter 1991
Lewis Porter, ed. *A Lester Young Reader*. Washington: Smithsonian Institution Press.

Priestley 1983
Brian Priestley. *Mingus: A Critical Biography*. New York: Quartet.

Reisner 1962
Robert George Reisner. *Bird: The Legend of Charlie Parker*. New York: Citadel.

Rosenthal 1992
David H. Rosenthal. *Hard Bop: Jazz and Black Music 1955–1965*. New York: Oxford.

Russell 1973
Ross Russell. *Bird Lives! The High Life and Hard Times of Charlie (Yardbird) Parker*. New York: Charterhouse.

Rust 1978
Brian Rust. *Jazz Records 1897–1942*. 4th ed. New Rochelle, N.Y.: Arlington. 2 vols.

Saks et al. 1989
Norman Saks, Leonard Bukowski, and Robert M. Bregman. *Yardbird: The Charlie Parker Discography*. N.p.: B.B.S.

Sancton 1990
Thomas Sancton. "Horns of Plenty," *Time* 136 (Oct. 22), 64ff.

Sawai n.d.
Genji Sawai, ed. *Immortal Jazz Series: John Coltrane*. Tokyo: Rittor.

Schuller 1958
Gunther Schuller. "Sonny Rollins and the Challenge of Thematic Improvisation," *Jazz Review* 1 (No. 1), 6ff.; reprinted in Williams 1964.

Schuller 1989
Gunther Schuller. *The Swing Era*. New York: Oxford.

Shapiro/Hentoff 1955
Nat Shapiro and Nat Hentoff. *Hear Me Talkin' to Ya*. New York: Rinehart.

Sickler 1986
Don Sickler. *John Coltrane Improvised Saxophone Solos*. Miami: Columbia Pictures.

Sjøgren 1983
Thorbjørn Sjøgren, compiler. *The Sonny Rollins Discography*. Copenhagen: Sjøgren.

Sjøgren 1986

Thorbjørn Sjøgren, compiler. *Long Tall Dexter: The Discography of Dexter Gordon.* Copenhagen: Sjøgren.

Slone 1982

Ken Slone. *Clifford Brown Trumpet Solos.* Louisville: Ducknob.

Slonimsky 1947

Nicolas Slonimsky. *Thesaurus of Scales and Melodic Patterns.* New York: Scribner's.

Spellman 1966

A. B. Spellman. *Four Lives in the Bebop Business.* New York: Pantheon.

Stewart 1974–75

Milton Stewart. "Structural Development in the Jazz Improvisational Techniques of Clifford Brown," *Jazzforschung* 6/7, pp. 141ff.

Taylor 1982

Billy Taylor. *Jazz Piano, a Jazz History.* Dubuque Iowa: Wm. C. Brown.

Tesser 1980

Neil Tesser. "Lee Konitz Searches for the Perfect Solo," *Down Beat* 47 (Jan.), 16ff.

Thomas 1975

J. C. Thomas. *Chasin' the Trane: The Music and Mystique of John Coltrane.* New York: Doubleday.

Tirro 1967

Frank Tirro. "The Silent Theme Tradition in Jazz," *Musical Quarterly* 53 (July), 313ff.

Velleman 1978

Barry Velleman. "Speaking of Jazz: Jazz Improvisation Through Linguistic Methods," *Music Educators Journal* 65 (Oct.), 28ff.

Watanabe n.d.

Sadao Watanabe, supervisor. *Jazz Improvisation: John Coltrane.* Tokyo: Nichion. 2 vols.

Weir 1986

Bob Weir. *Clifford Brown Discography,* 3rd ed. Wales: Weir.

Wernboe 1986

Roger Wernboe. *Lee Morgan Discography.* 2nd ed. Saltsjobaden, Sweden: Wernboe.

Werther 1988

Iron Werther. *Bebop: Die Geschichte einer musikalischen Revolution und ihrer Interpreten.* Frankfurt am Main: Fischer.

Wild 1979

David Wild. *The Recordings of John Coltrane: A Discography.* 2nd ed. Ann Arbor: Wild Music.

Williams 1964

Martin Williams, ed. *Jazz Panorama.* New York: Collier.

Williams 1970

Martin Williams. *The Jazz Tradition.* New York: Oxford.

Woodson 1973
Craig DeVere Woodson. *Solo Jazz Drumming: An Analytic Study of the Improvisation Techniques of Anthony Williams.* M.A. thesis, University of California at Los Angeles. 2 vols.

Yoshida n.d.
Kenji Yoshida, ed. *Immortal Jazz Series: Clifford Brown* [transcriptions]. Tokyo: Intersong.

Index